The Great Books Reading & Discussion Program

FIRST SERIES • VOLUME TWO

The Great Books Foundation
A Nonprofit Educational Corporation

Designed by Don Walkoe Design, Chicago

Handmade marbled paper, photographed on cover, courtesy of Andrews, Nelson, Whitehead.

SFI® Certified Sourcing
www.sfiprogram.org
SFI-00453

ISBN 0-945159-77-3

15 14 13 12 11

Published and distributed by

The Great Books Foundation
A Nonprofit Educational Corporation
35 East Wacker Drive, Suite 400
Chicago, IL 60601

Acknowledgments

"Rothschild's Fiddle" from *The Oxford Chekhov, Vol. VII: Stories 1893–1895,* translated and edited by Ronald Hingley. Copyright 1978 by Ronald Hingley. Reprinted by permission of the publisher, Oxford University Press.

"On Happiness" from *Nichomachean Ethics* by Aristotle, translated by Martin Ostwald. Copyright 1962 by The Bobbs-Merrill Company, Inc. Reprinted by permission of the publisher, The Bobbs-Merrill Company, Inc.

"The Apology" by Plato from *Socrates and Legal Obligation,* translated by R. E. Allen. Copyright 1980 by the University of Minnesota. Reprinted by permission of the publisher, The University of Minnesota Press.

"Conscience" from *Lectures on Ethics* by Immanuel Kant, translated by Louis Infield. Published by Methuen & Co., Ltd. Reprinted by permission of Associated Book Publishers, Ltd.

"Alienated Labour" from *Karl Marx, Early Writings,* translated and edited by T. B. Bottomore. Copyright 1963 by T. B. Bottomore. Reprinted by permission of the publisher, McGraw-Hill Book Company.

"Civilization and Its Discontents" from *Civilization and Its Discontents,* translated by James Strachey. Copyright 1961 by James Strachey. Reprinted by permission of the publisher, W. W. Norton & Company, Inc.

"The Social Contract" from *On the Social Contract* by Jean-Jacques Rousseau, translated by Judith R. Masters, edited by Roger D. Masters. Copyright 1978 by St. Martin's Press, Inc. Reprinted by permission of the publisher, St. Martin's Press, Inc.

"Of Justice and Injustice" from *Hume's Moral and Political Philosophy,* edited by Henry D. Aiken. Copyright 1948 by Hafner Press, a division of Macmillan Publishing Company, Inc. Reprinted by permission of the publisher, Macmillan Publishing Company, Inc.

"Individual Freedom" from *The Philosophy of Money* by Georg Simmel, translated by Tom Bottomore and David Frisby. Copyright 1978 by Routledge and Kegan Paul, Ltd. Reprinted by permission of the publisher, Routledge and Kegan Paul, Ltd.

"Antigone" from *The Antigone of Sophocles:* An English Version by Dudley Fitts and Robert Fitzgerald. Copyright 1967 by Dudley Fitts and Robert Fitzgerald. Reprinted by permission of the publisher, Harcourt Brace Jovanovich, Inc.

A source note appears, together with biographical information about the author, opposite the opening page of each work in this series. Footnotes by the author are not bracketed; footnotes by GBF or a translator are [bracketed].

CONTENTS

*

CHARLES DARWIN was born in 1809 in Shrewsbury, England, descended from scientist Erasmus Darwin and potter Josiah Wedgwood. As a young man, Darwin studied medicine and theology without enthusiasm or success. Complained his father: "You care for nothing but shooting, dogs, and rat-catching, and you will be a disgrace to yourself and all your family." In 1831 Darwin joined the crew of a surveying expedition as a naturalist, sailing on H.M.S. *Beagle* to South America, Australia, South Africa, and islands in the Pacific over a period of five years. Darwin's observations in the field formed the foundation of his theories of natural selection and evolution that were first presented in *The Origin of Species* (1859). This work and others brought Darwin under attack by scientists and churchmen; his theories challenged the conclusions of previous research and beliefs basic to Christian doctrine. Darwin died in 1881 after years of illness caused by insect bites suffered during the *Beagle* expedition.

From *The Origin of Species and The Descent of Man.*
Publisher: The Modern Library, Random House, Inc.
Pages 471–95.

The Moral Sense of Man and the Lower Animals

I fully subscribe to the judgment of those writers who maintain that of all the differences between man and the lower animals, the moral sense or conscience is by far the most important. This sense, as Mackintosh remarks, "has a rightful supremacy over every other principle of human action"; it is summed up in that short but imperious word *ought,* so full of high significance. It is the most noble of all the attributes of man, leading him without a moment's hesitation to risk his life for that of a fellow creature; or after due deliberation, impelled simply by the deep feeling of right or duty, to sacrifice it in some great cause. Immanuel Kant exclaims, "Duty! Wondrous thought, that workest neither by fond insinuation, flattery, nor by any threat, but merely by holding up thy naked law in the soul, and so extorting for thyself always reverence, if not always obedience; before whom all appetites are dumb, however secretly they rebel; whence thy original?"

This great question has been discussed by many writers of consummate ability; and my sole excuse for touching on it, is the impossibility of here passing it over; and because, as far as I know, no one has approached it exclusively from the side of natural history. The investigation possesses, also, some independent interest, as an attempt to see how far the study of the lower animals throws light on one of the highest physical faculties of man.

The following proposition seems to me in a high degree probable—namely, that any animal whatever, endowed with well-marked social instincts, the parental and filial affections being here included, would inevitably acquire a moral sense or conscience, as soon as its intellectual powers had become as well, or nearly as well developed, as in man. For, firstly, the social instincts lead an animal to take pleasure in the society of its fellows, to feel a certain amount of sympathy with them, and to perform various services for them. The services may be of a definite and evidently instinctive nature; or there may be only a wish and readiness, as with most of the higher social animals, to aid their fellows in certain general ways. But these feelings and services are by no means extended to all the individuals of the same species, only to those of the same association. Secondly, as soon as the mental faculties had become highly developed, images of all past actions and motives would be incessantly passing through the brain of each individual: and that feeling of dissatisfaction, or even misery, which invariably results, as we shall hereafter see, from any unsatisfied instinct, would arise as often as it was perceived that the enduring and always present social instinct had yielded to some other instinct, at the time stronger, but neither enduring in its nature, nor leaving behind it a very vivid impression. It is clear that many instinctive desires, such as that of hunger, are in their nature of short duration; and after being satisfied, are not readily or vividly recalled. Thirdly, after the power of language had been acquired, and the wishes of the community could be expressed, the common opinion how each member ought to act for the public good would naturally become in a paramount degree the guide to action. But it should be borne in mind that however great weight we may attribute to public opinion, our regard for the approbation and disapprobation of our fellows depends on sympathy, which, as we shall see, forms an essential part of the social instinct, and is indeed its foundation stone. Lastly, habit in the individual would ultimately play a very important part in guiding the conduct of each member; for the social instinct, together with sympathy, is, like any other instinct, greatly strengthened by habit, and so consequently would be obedience to the wishes and judgment of the community. These several subordinate propositions must now be discussed, and some of them at considerable length.

It may be well first to premise that I do not wish to maintain that any strictly social animal, if its intellectual faculties were to become as active and as highly developed as in man, would acquire exactly the same moral sense as ours. In the same manner as various animals have some sense of beauty, though they admire widely different objects, so they might have a sense of right and wrong, though led by it to follow widely different lines of conduct. If, for instance, to take an extreme case, men were reared under precisely the same conditions as hive-bees, there can hardly be a doubt that our unmarried females would, like the worker-bees, think it a sacred duty to kill their brothers, and mothers would strive to kill their fertile daughters; and no one would think of interfering. Nevertheless, the bee, or any other social animal, would gain in our supposed case, as it appears to me, some feeling of right or wrong, or a conscience. For each individual would have an inward sense of possessing certain stronger or more enduring instincts, and others less strong or enduring; so that there would often be a struggle as to which impulse should be followed; and satisfaction, dissatisfaction, or even misery would be felt, as past impressions were compared during their incessant passage through the mind. In this case an inward monitor would tell the animal that it would have been better to have followed the one impulse rather than the other. The one course ought to have been followed, and the other ought not; the one would have been right and the other wrong; but to these terms I shall recur.

Sociability—Animals of many kinds are social; we find even distinct species living together; for example, some American monkeys, and united flocks of rooks, jackdaws, and starlings. Man shows the same feeling in his strong love for the dog, which the dog returns with interest. Everyone must have noticed how miserable horses, dogs, sheep, etc., are when separated from their companions, and what strong mutual affection the two former kinds, at least, show on their reunion. It is curious to speculate on the feelings of a dog, who will rest peacefully for hours in a room with his master or any of the family, without the least notice being taken of him; but if left for a short time by himself, barks or howls dismally. We will confine our attention to the higher social animals, and pass over insects, although some of these are social, and aid one another in many important ways. The most common mutual service in the higher animals is to warn one another of danger by means of the united senses of all. Every sportsman knows, as Dr. Jaeger remarks, how difficult it is to approach animals in a herd or troop. Wild horses and cattle do not, I believe, make any danger signal; but the attitude of any one of them who first discovers an enemy warns the others. Rabbits stamp loudly on the ground with their hind feet as a signal: sheep and chamois do the same with their forefeet, uttering likewise a whistle. Many birds, and some mammals, post sentinels, which in the case of seals are said generally to be the females. The leader of a troop of monkeys acts as the sentinel, and utters cries expressive both of danger and of safety. Social animals perform many little services for each other: horses nibble, and cows lick each other, on any spot which itches; monkeys search each other for external parasites; and Brehm states that after a troop of the *Cercopithecus griseoviridis* has rushed through a thorny brake, each monkey stretches itself on a branch, and another monkey sitting by "conscientiously" examines its fur, and extracts every thorn or burr.

Animals also render more important services to one another: thus wolves and some other beasts of prey hunt in packs, and aid one another in attacking their victims. Pelicans fish in concert.

The Hamadryas baboons turn over stones to find insects, etc., and when they come to a large one, as many as can stand round turn it over together and share the booty. Social animals mutually defend each other. Bull bisons in North America, when there is danger, drive the cows and calves into the middle of the herd, while they defend the outside. I shall also in a future chapter give an account of two young wild bulls at Chillingham attacking an old one in concert, and of two stallions together trying to drive away a third stallion from a troop of mares. In Abyssinia, Brehm encountered a great troop of baboons who were crossing a valley: some had already ascended the opposite mountain, and some were still in the valley; the latter were attacked by the dogs, but the old males immediately hurried down from the rocks, and with mouths widely opened, roared so fearfully that the dogs quickly drew back. They were again encouraged to the attack; but by this time all the baboons had reascended the heights, excepting a young one, about six months old, who, loudly calling for aid, climbed on a block of rock, and was surrounded. Now one of the largest males, a true hero, came down again from the mountain, slowly went to the young one, coaxed him, and triumphantly led him away—the dogs being too much astonished to make an attack. I cannot resist giving another scene which was witnessed by this same naturalist: an eagle seized a young Cercopithecus, which, by clinging to a branch, was not at once carried off; it cried loudly for assistance, upon which the other members of the troop, with much uproar, rushed to the rescue, surrounded the eagle, and pulled out so many feathers that he no longer thought of his prey, but only how to escape. This eagle, as Brehm remarks, assuredly would never again attack a single monkey of a troop.

It is certain that associated animals have a feeling of love for each other, which is not felt by non-social adult animals. How far in most cases they actually sympathise in the pains and pleasures of others is more doubtful, especially with respect to pleasures. Mr. Buxton, however, who had excellent means of observation, states that his macaws, which lived free in Norfolk, took "an extravagant interest" in a pair with a nest; and whenever the female left it, she was surrounded by a troop "screaming horrible acclamations in her honour." It is often difficult to judge whether animals have any feeling for the sufferings of others of their kind. Who can say what cows feel when they surround and stare intently on a dying or dead companion; apparently, however, as Houzeau remarks, they feel no pity. That animals sometimes are far from feeling any sympathy is too certain; for they will expel a wounded animal from the herd, or gore or worry it to death. This is almost the blackest fact in natural history, unless, indeed, the explanation which has been suggested is true, that their instinct or reason leads them to expel an injured companion, lest beasts of prey, including man, should be tempted to follow the troop. In this case their conduct is not much worse than that of the North American Indians, who leave their feeble comrades to perish on the plains; or the Fijians, who, when their parents get old, or fall ill, bury them alive.

Many animals, however, certainly sympathise with each other's distress or danger. This is the case even with birds. Captain Stansbury found on a salt lake in Utah an old and completely blind pelican, which was very fat, and must have been well fed for a long time by his companions. Mr. Blyth, as he informs me, saw Indian crows feeding two or three of their companions which were blind; and I have heard of an analogous case with the domestic cock. We may, if we choose, call these actions instinctive; but such cases are much too rare for the development of any special instinct. I have myself seen a dog who never passed a cat who lay sick in a basket, and was a great friend of his, without giving her a few licks with his tongue, the surest sign of kind feeling in a dog.

It must be called sympathy that leads a courageous dog to fly at anyone who strikes his master, as he certainly will. I saw a person pretending to beat a lady, who had a very timid little dog on her lap, and the trial had never been made before; the little creature instantly jumped away, but after the pretended beating was over, it was really pathetic to see how perseveringly he tried to lick his mistress' face, and comfort her. Brehm states that when a baboon in confinement was pursued to be punished, the others tried to protect him. It must have been sympathy in the cases above which led the baboons and Cercopitheci to defend their young comrades from the dogs and the eagle. I will give only one other instance of sympathetic and heroic conduct, in the case of a little American monkey. Several years ago a keeper at the Zoological Gardens showed me some deep and scarcely healed wounds on the nape of his own neck, inflicted on him, while kneeling on the floor, by a fierce baboon. The little American monkey, who was a warm friend of this keeper, lived in the same compartment, and was dreadfully afraid of the great baboon. Nevertheless, as soon as he saw his friend in peril, he rushed to the rescue, and by screams and bites so distracted the baboon that the man was able to escape, after, as the surgeon thought, running great risk of his life.

Besides love and sympathy, animals exhibit other qualities connected with the social instincts, which in us would be called moral; and I agree with Agassiz that dogs possess something very like a conscience.

Dogs possess some power of self-command, and this does not appear to be wholly the result of fear. As Braubach remarks, they will refrain from stealing food in the absence of their master. They have long been accepted as the very type of fidelity and obedience. But the elephant is likewise very faithful to his driver or keeper, and probably considers him as the leader of the herd. Dr. Hooker informs me that an elephant, which he was riding in India, became so deeply bogged that he remained stuck fast until the next day, when he was extricated by men with ropes. Under such circumstances elephants will seize with their trunks any object, dead or alive, to place under their knees, to prevent their sinking deeper in the mud; and the driver was dreadfully afraid lest the animal should have seized Dr. Hooker and crushed him to death. But the driver himself, as Dr. Hooker was assured, ran no risk. This forbearance under an emergency so dreadful for a heavy animal is a wonderful proof of noble fidelity.

All animals living in a body, which defend themselves or attack their enemies in concert, must indeed be in some degree faithful to one another; and those that follow a leader must be in some degree obedient. When the baboons in Abyssinia plunder a garden, they silently follow their leader; and if an imprudent young animal makes a noise, he receives a slap from the others to teach him silence and obedience. Mr. Galton, who has had excellent opportunities for observing the half-wild cattle in South Africa, says that they cannot endure even a momentary separation from the herd. They are essentially slavish, and accept the common determination, seeking no better lot than to be led by any one ox who has enough self-reliance to accept the position. The men who break in these animals for harness watch assiduously for those who, by grazing apart, show a self-reliant disposition, and these they train as fore-oxen. Mr. Galton adds that such animals are rare and valuable; and if many were born they would soon be eliminated, as lions are always on the look-out for the individuals which wander from the herd.

With respect to the impulse which leads certain animals to associate together, and to aid one another in many ways, we may infer that in most cases they are impelled by the same sense of satisfaction or pleasure which they experience in performing other instinctive actions; or by the same sense of dissatisfaction as when other instinctive actions are checked. We see this in innumerable instances, and it is illustrated in a striking manner by the acquired instincts of our domesticated animals; thus a young shepherd dog delights in driving and running round a flock of sheep, but not in worrying them; a young foxhound delights in hunting a fox, while some other kinds of dogs, as I have witnessed, utterly disregard foxes. What a strong feeling of inward satisfaction must impel a bird, so full of activity, to brood day after day over her eggs. Migratory birds are quite miserable if stopped from migrating—perhaps they enjoy starting on their long flight; but it is hard to believe that the poor pinioned goose, described by Audubon, which started on foot at the proper time for its journey of probably more than a thousand miles, could have felt any joy in doing so. Some instincts are determined solely by painful feelings, as by fear, which leads to self-preservation, and is in some cases directed towards special enemies. No one, I presume, can analyse the sensations of pleasure or pain. In many instances, however, it is probable that instincts are persistently followed from the mere force of inheritance, without the stimulus of either pleasure or pain. A young pointer, when it first scents game, apparently cannot help pointing. A squirrel in a cage who pats the nuts which it cannot eat, as if to bury them in the ground, can hardly be thought to act thus either from pleasure or pain. Hence the common assumption that men must be impelled to every action by experiencing some pleasure or pain may be erroneous. Although a habit may be blindly and implicitly followed, independently of any pleasure or pain felt at the moment, yet if it be forcibly and abruptly checked, a vague sense of dissatisfaction is generally experienced.

It has often been assumed that animals were in the first place rendered social, and that they feel as a consequence uncomfortable when separated from each other, and comfortable while together; but it is a more probable view that these sensations were first developed in order that those animals which would profit by living in society should be induced to live together, in the same manner as the sense of hunger and the pleasure of eating were, no doubt, first acquired in order to induce animals to eat. The feeling of pleasure from society is probably an extension of the parental or filial affections, since the social instinct seems to be developed by the young remaining for a long time with their parents; and this extension may be attributed in part to habit, but chiefly to natural selection.[1] With those animals which were benefited by living in close association, the individuals which took the greatest pleasure in society would best escape various dangers, while those that cared least for their comrades, and lived solitary, would perish in greater numbers. With respect to the origin of the parental and filial affections, which apparently lie at the base of the social instincts, we know not the steps by which they have been gained, but we may infer that it has been to a large extent through natural selection. So it has almost certainly been with the unusual and opposite feeling of hatred between the nearest relations, as with the worker-bees which kill their brother drones, and with the queen-bees which kill their daughter-queens; the desire to destroy their nearest relations having been in this case of service to the community. Parental affection, or some feeling which replaces it, has been developed in certain animals extremely low in the scale, for example, in starfishes and spiders. It is also occasionally present in a few members alone in a whole group of animals, as in the genus Forficula, or earwigs.

The all-important emotion of sympathy is distinct from that of love. A mother may passionately love her sleeping and passive infant, but she can hardly at such times be said to feel sympathy for it. The love of a man for his dog is distinct from sympathy, and so is that of a dog for his master. Adam Smith formerly argued, as has Mr. Bain recently, that the basis of sympathy

---

[1] [Darwin defines natural selection as the tendency in individuals and species for variations that are favorable for survival to be preserved in the struggle for existence, and for injurious variations to be eliminated. He considers natural selection a major cause of the modification and development of life forms.]

lies in our strong retentiveness of former states of pain or pleasure. Hence, "the sight of another person enduring hunger, cold, fatigue, revives in us some recollection of these states, which are painful even in idea." We are thus impelled to relieve the sufferings of another, in order that our own painful feelings may be at the same time relieved. In like manner we are led to participate in the pleasures of others. But I cannot see how this view explains the fact that sympathy is excited, in an immeasurably stronger degree, by a beloved, than by an indifferent person. The mere sight of suffering, independently of love, would suffice to call up in us vivid recollections and associations. The explanation may lie in the fact that, with all animals, sympathy is directed solely towards the members of the same community, and therefore towards known, and more or less beloved members, but not to all the individuals of the same species. This fact is not more surprising than that the fears of many animals should be directed against special enemies. Species which are not social, such as lions and tigers, no doubt feel sympathy for the suffering of their own young, but not for that of any other animal. With mankind, selfishness, experience, and imitation, probably add, as Mr. Bain has shown, to the power of sympathy; for we are led by the hope of receiving good in return to perform acts of sympathetic kindness to others; and sympathy is much strengthened by habit. In however complex a manner this feeling may have originated, as it is one of high importance to all those animals which aid and defend one another, it will have been increased through natural selection; for those communities, which included the greatest number of the most sympathetic members, would flourish best, and rear the greatest number of offspring.

It is, however, impossible to decide in many cases whether certain social instincts have been acquired through natural selection, or are the indirect result of other instincts and faculties, such as sympathy, reason, experience, and a tendency to imitation; or again, whether they are simply the result of long-continued habit. So remarkable an instinct as the placing sentinels to warn the community of danger, can hardly have been the indirect result of any of these faculties; it must, therefore, have been directly acquired. On the other hand, the habit followed by the males of some social animals of defending the community, and of attacking their enemies or their prey in concert, may perhaps have originated from mutual sympathy; but courage, and in most cases strength, must have been previously acquired, probably through natural selection.

Of the various instincts and habits, some are much stronger than others; that is, some either give more pleasure in their performance, and more distress in their prevention, than others; or, which is probably quite as important, they are, through inheritance, more persistently followed, without exciting any special feeling of pleasure or pain. We are ourselves conscious that some habits are much more difficult to cure or change than others. Hence a struggle may often be observed in animals between different instincts, or between an instinct and some habitual disposition; as when a dog rushes after a hare, is rebuked, pauses, hesitates, pursues again, or returns ashamed to his master; or as between the love of a female dog for her young puppies and for her master—for she may be seen to slink away to them, as if half ashamed of not accompanying her master. But the most curious instance known to me of one instinct getting the better of another, is the migratory instinct conquering the maternal instinct. The former is wonderfully strong; a confined bird will at the proper season beat her breast against the wires of her cage, until it is bare and bloody. It causes young salmon to leap out of the fresh water, in which they could continue to exist, and thus unintentionally to commit suicide. Everyone knows how strong the maternal instinct is, leading even timid birds to face great danger, though with hesitation, and in opposition to the instinct of self-preservation. Nevertheless, the migratory instinct is so powerful, that late in the autumn swallows, housemartins, and swifts frequently desert their tender young, leaving them to perish miserably in their nests.

We can perceive that an instinctive impulse, if it be in any way more beneficial to a species than some other or opposed instinct, would be rendered the more potent of the two through natural selection; for the individuals which had it most strongly developed would survive in larger numbers. Whether this is the case with the migratory in comparison with the maternal instinct, may be doubted. The great persistence, or steady action of the former at certain seasons of the year during the whole day, may give it for a time paramount force.

Man a Social Animal—Everyone will admit that man is a social being. We see this in his dislike of solitude, and in his wish for society beyond that of his own family. Solitary confinement is one of the severest punishments which can be inflicted. Some authors suppose that man primevally lived in single families; but at the present day, though single families, or only two or three together, roam the solitudes of some savage lands, they always, as far as I can discover, hold friendly relations with other families inhabiting the same district. Such families occasionally meet in council, and unite for their common defence. It is no argument against savage man being a social animal that the tribes inhabiting adjacent districts are almost always at war with each other; for the social instincts never extend to all the individuals of the same species. Judging from the analogy of the majority of the Quadrumana, it is probable that the early apelike progenitors of man were likewise social; but this is not of much importance for us. Although man, as he now exists, has few special instincts, having lost any which his early progenitors may have possessed, this is no reason why he should not have retained from an extremely remote period some degree of instinctive love and sympathy for his fellows. We are indeed all conscious that we do possess such sympathetic feelings; but our consciousness does not tell us whether they are instinctive, having originated long ago in the same manner as with the lower animals, or whether they have been acquired by each of us during our early years. As man is a social animal, it is almost certain that he would inherit a tendency to be faithful to his comrades, and obedient to the leader of his tribe; for these qualities are common to most social animals. He would consequently possess some capacity for self-command. He would from an inherited tendency be willing to defend, in concert with others, his fellow men; and would be ready to aid them in any way which did not too greatly interfere with his own welfare or his own strong desires.

The social animals which stand at the bottom of the scale are guided almost exclusively, and those which stand higher in the scale are largely guided, by special instincts in the aid which they give to the members of the same community; but they are likewise in part impelled by mutual love and sympathy, assisted apparently by some amount of reason. Although man, as just remarked, has no special instincts to tell him how to aid his fellow men, he still has the impulse, and with his improved intellectual faculties would naturally be much guided in this respect by reason and experience. Instinctive sympathy would also cause him to value highly the approbation of his fellows; for, as Mr. Bain has clearly shown, the love of praise and the strong feeling of glory, and the still stronger horror of scorn and infamy, "are due to the workings of sympathy." Consequently man would be influenced in the highest degree by the wishes, approbation, and blame of his fellow men, as expressed by their gestures and language. Thus the social instincts, which must have been acquired by man in a very rude state, and probably even by his early ape-like progenitors, still give the impulse to some of his best actions; but his actions are in a higher degree determined by the expressed wishes and judgment of his fellow men, and unfortunately very often by his own strong selfish desires. But as love, sympathy, and self-command become strengthened by habit, and as the power of reasoning becomes clearer, so that man can value justly the judgments of his fellows, he will feel himself impelled, apart from any transitory pleasure or pain, to certain lines of conduct. He might then declare— not that any barbarian or uncultivated man could thus think— I am the supreme judge of my own conduct, and in the words of Kant, I will not in my own person violate the dignity of humanity.

The More Enduring Social Instincts Conquer the Less Persistent Instincts—We have not, however, as yet considered the main point, on which, from our present point of view, the whole question of the moral sense turns. Why should a man feel that he ought to obey one instinctive desire rather than another? Why is he bitterly regretful, if he has yielded to a strong sense of self-preservation, and has not risked his life to save that of a fellow creature? Or why does he regret having stolen food from hunger?

It is evident in the first place, that with mankind the instinctive impulses have different degrees of strength; a savage will risk his own life to save that of a member of the same community, but will be wholly indifferent about a stranger; a young and timid mother urged by the maternal instinct will, without a moment's hesitation, run the greatest danger for her own infant, but not for a mere fellow creature. Nevertheless many a civilized man, or even boy, who never before risked his life for another, but full of courage and sympathy, has disregarded the instinct of self-preservation, and plunged at once into a torrent to save a drowning man, though a stranger. In this case man is impelled by the same instinctive motive which made the heroic little American monkey, formerly described, save his keeper, by attacking the great and dreaded baboon. Such actions as the above appear to be the simple result of the great strength of the social or maternal instincts than that of any other instinct or motive; for they are performed too instantaneously for reflection, or for pleasure or pain to be felt at the time; though, if prevented by any cause, distress or even misery might be felt. In a timid man, on the other hand, the instinct of self-preservation might be so strong, that he would be unable to force himself to run any such risk, perhaps not even for his own child.

I am aware that some persons maintain that actions performed impulsively, as in the above cases, do not come under the dominion of the moral sense, and cannot be called moral. They confine this term to actions done deliberately, after a victory over opposing desires, or when prompted by some exalted motive. But it appears scarcely possible to draw any clear line of distinction of this kind. As far as exalted motives are concerned, many instances have been recorded of savages, destitute of any feeling of general benevolence towards mankind, and not guided by any religious motive, who have deliberately sacrificed their lives as prisoners, rather then betray their comrades; and surely their conduct ought to be considered as moral. As far as deliberation, and the victory over opposing motives are concerned, animals may be seen doubting between opposed instincts in rescuing their offspring or comrades from danger; yet their actions, though done for the good of others, are not called moral. Moreover, anything performed very often by us will at last be done without deliberation or hesitation, and can then hardly be distinguished from an instinct; yet surely no one will pretend that such an action ceases to be moral. On the contrary, we all feel that an act cannot be considered as perfect, or as performed in the most noble manner, unless it be done impulsively, without deliberation or effort, in the same manner as by a man in whom the requisite qualities are innate. He who is forced to overcome his fear or want of sympathy before he acts, deserves, however, in one way higher credit than the man whose innate disposition leads him to a good act without effort. As we cannot distinguish between motives, we rank all actions of a certain class as moral, if performed by a moral being. A moral being is one who is capable of comparing his past and future actions or motives, and of approving or disapproving of them. We have no reason to suppose that any of the lower animals have this capacity; therefore, when a Newfoundland dog drags a child out of the water, or a monkey faces danger to rescue its comrade, or takes charge of an orphan monkey, we do not call its conduct moral. But in the case of man, who alone can with certainty be ranked as a moral being, actions of a certain class are called moral, whether performed deliberately, after a struggle with opposing motives, or impulsively through instinct, or from the effects of slowly gained habit.

But to return to our more immediate subject. Although some instincts are more powerful than others, and thus lead to corresponding actions, yet it is untenable that in man the social instincts (including the love of praise and fear of blame) possess greater strength, or have, through long habit, acquired greater strength than the instincts of self-preservation, hunger, lust, vengeance, etc. Why then does man regret, even though trying to banish such regret, that he has followed the one natural impulse rather than the other; and why does he further feel that he ought to regret his conduct? Man in this respect differs profoundly from the lower animals. Nevertheless we can, I think, see with some degree of clearness the reason of this difference.

Man, from the activity of his mental faculties, cannot avoid reflection: past impressions and images are incessantly and clearly passing through his mind. Now with those animals which live permanently in a body, the social instincts are ever present and persistent. Such animals are always ready to utter the danger-signal, to defend the community, and to give aid to their fellows in accordance with their habits; they feel at all times, without the stimulus of any special passion or desire, some degree of love and sympathy for them; they are unhappy if long separated from them, and always happy to be again in their company. So it is with ourselves. Even when we are quite alone, how often do we think with pleasure or pain of what others think of us —of their imagined approbation or disapprobation; and this all follows from sympathy, a fundamental element of the social instincts. A man who possessed no trace of such instincts would be an unnatural monster. On the other hand, the desire to satisfy hunger, or any passion such as vengeance, is in its nature temporary, and can for a time be fully satisfied. Nor is it easy, perhaps hardly possible, to call up with complete vividness the feeling, for instance, of hunger; nor indeed, as has often been remarked, of any suffering. The instinct of self-preservation is not felt except in the presence of danger; and many a coward has thought himself brave until he has met his enemy face to face. The wish for another man's property is perhaps as persistent a desire as any that can be named; but even in this case the satisfaction of actual possession is generally a weaker feeling than the desire: many a thief, if not a habitual one, after success has wondered why he stole some article.[2]

A man cannot prevent past impressions often repassing through his mind; he will thus be driven to make a comparison between the impressions of past hunger, vengeance satisfied, or danger

---

[2] Enmity or hatred seems also to be a highly persistent feeling, perhaps more so than any other that can be named. Envy is defined as hatred of another for some excellence or success; and Bacon insists, "Of all other affections envy is the most importune and continual." Dogs are very apt to hate both strange men and strange dogs, especially if they live near at hand, but do not belong to the same family, tribe, or clan; this feeling would thus seem to be innate, and is certainly a most persistent one. It seems to be the complement and converse of the true social instinct. From what we hear of savages, it would appear that something of the same kind holds good with them. . . . To do good in return for evil, to love your enemy, is a height of morality to which it may be doubted whether the social instincts would, by themselves, have ever led us. It is necessary that these instincts, together with sympathy, should have been highly cultivated and extended by the aid of reason, instruction, and the love or fear of God, before any such golden rule would ever be thought of and obeyed.

shunned at other men's cost, with the almost ever-present instinct of sympathy, and with his early knowledge of what others consider as praiseworthy or blameable. This knowledge cannot be banished from his mind, and from instinctive sympathy is esteemed of great moment. He will then feel as if he had been baulked in following a present instinct or habit, and this with all animals causes dissatisfaction, or even misery.

The above case of the swallow affords an illustration, though of a reversed nature, of a temporary though for the time strongly persistent instinct conquering another instinct, which is usually dominant over all others. At the proper season these birds seem all day long to be impressed with the desire to migrate; their habits change; they become restless, are noisy, and congregate in flocks. While the mother bird is feeding, or brooding over her nestlings, the maternal instinct is probably stronger than the migratory; but the instinct which is the more persistent gains the victory, and at last, at a moment when her young ones are not in sight, she takes flight and deserts them. When arrived at the end of her long journey, and the migratory instinct has ceased to act, what an agony of remorse the bird would feel, if, from being endowed with great mental activity, she could not prevent the image constantly passing through her mind, of her young ones perishing in the bleak north from cold and hunger.

At the moment of action, man will no doubt be apt to follow the stronger impulse; and though this may occasionally prompt him to the noblest deeds, it will more commonly lead him to gratify his own desires at the expense of other men. But after their gratification when past and weaker impressions are judged by the ever-enduring social instinct, and by his deep regard for the good opinion of his fellows, retribution will surely come. He will then feel remorse, repentance, regret, or shame; this latter feeling, however, relates almost exclusively to the judgment of others. He will consequently resolve more or less firmly to act differently for the future; and this is conscience; for conscience looks backwards, and serves as a guide for the future.

The nature and strength of the feelings which we call regret, shame, repentance, or remorse, depend apparently not only on the strength of the violated instinct, but partly on the strength of the temptation, and often still more on the judgment of our fellows. How far each man values the appreciation of others depends on the strength of his innate or acquired feeling of sympathy, and on his own capacity for reasoning out the remote consequences of his acts. Another element is most important, although not necessary: the reverence or fear of the Gods, or Spirits believed in by each man; and this applies especially in cases of remorse. Several critics have objected that though some slight regret or repentance may be explained by the view advocated in this chapter, it is impossible thus to account for the soul-shaking feeling of remorse. But I can see little force in this objection. My critics do not define what they mean by remorse, and I can find no definition implying more than an overwhelming sense of repentance. Remorse seems to bear the same relation to repentance, as rage does to anger, or agony to pain. It is far from strange that an instinct so strong and so generally admired as maternal love, should, if disobeyed, lead to the deepest misery, as soon as the impression of the past cause of disobedience is weakened. Even when an action is opposed to no special instinct, merely to know that our friends and equals despise us for it is enough to cause great misery. Who can doubt that the refusal to fight a duel through fear has caused many men an agony of shame? Many a Hindoo, it is said, has been stirred to the bottom of his soul by having partaken of unclean food. Here is another case of what must, I think, be called remorse. Dr. Landor acted as a magistrate in West Australia, and relates that a native on his farm, after losing one of his wives from disease, came and said that "he was going to a distant tribe to spear a woman, to satisfy his sense of duty to his wife. I told him that if he did so, I would send him to prison for life. He remained about the farm for some months, but got exceedingly thin, and complained that he could not rest or eat, that his wife's spirit was haunting him, because he had not taken a life for hers. I was inexorable, and assured him that nothing should save him if he did." Nevertheless the man disappeared for more than a year, and then returned in high condition; and his other wife told Dr. Landor that her husband had taken the life of a woman belonging to a distant tribe; but it was impossible to obtain legal evidence of the act. The breach of a rule held sacred by the tribe will thus, as it seems, give rise to the deepest feelings—and this quite apart from the social instincts, excepting insofar as the rule is grounded on the judgment of the community. How so many strange superstitions have arisen throughout the world we know not; nor can we tell how some real and great crimes, such as incest, have come to be held in an abhorrence (which is not, however, quite universal) by the lowest savages. It is even doubtful whether in some tribes incest would be looked on with greater horror than would the marriage of a man with a woman bearing the same name, though not a relation. "To violate this law is a crime which the Australians hold in the greatest abhorrence, in this agreeing exactly with certain tribes of North America. When the question is put in either district, is it worse to kill a girl of a foreign tribe, or to marry a girl of one's own, an answer just opposite to ours would be given without hesitation."[3] We may, therefore, reject the belief, lately insisted on by some writers, that the abhorrence of incest is due to our possessing a special God-implanted conscience. On the whole it is intelligible that a man urged by so powerful a sentiment as remorse, though arising as above explained, should be led to act in a manner which he has been taught to believe serves as an expiation, such as delivering himself up to justice.

---

[3] E. B. Taylor in *Contemporary Review,* April, 1873, p. 707.

Man, prompted by his conscience, will through long habit acquire such perfect self-command that his desires and passions will at last yield instantly and without a struggle to his social sympathies and instincts, including his feeling for the judgment of his fellows. The still hungry, or the still revengeful man will not think of stealing food, or of wreaking his vengeance. It is possible, or as we shall hereafter see, even probable, that the habit of self-command may, like other habits, be inherited. Thus at last man comes to feel, through acquired and perhaps inherited habit, that it is best for him to obey his more persistent impulses. The imperious word *ought* seems merely to imply the consciousness of the existence of a rule of conduct, however it may have originated. Formerly it must have been often vehemently urged that an insulted gentleman *ought* to fight a duel. We even say that a pointer *ought* to point, and a retriever to retrieve game. If they fail to do so, they fail in their duty and act wrongly.

If any desire or instinct leading to an action opposed to the good of others still appears, when recalled to mind, as strong as, or stronger than, the social instinct, a man will feel no keen regret at having followed it; but he will be conscious that if his conduct were known to his fellows, it would meet with their disapprobation; and few are so destitute of sympathy as not to feel discomfort when this is realised. If he has no such sympathy, and if his desires leading to bad actions are at the time strong, and when recalled are not over-mastered by the persistent social instincts, and the judgment of others, then he is essentially a bad man; and the sole restraining motive left is the fear of punishment, and the conviction that in the long run it would be best for his own selfish interests to regard the good of others rather than his own.

It is obvious that everyone may with an easy conscience gratify his own desires if they do not interfere with his social instincts, that is, with the good of others; but in order to be quite free from self-reproach, or at least of anxiety, it is almost necessary for him to avoid the disapprobation, whether reasonable or not, of his fellow men. Nor must be break through the fixed habits of his life, especially if these are supported by reason; for if he does, he will assuredly feel dissatisfaction. He must likewise avoid the reprobation of the one God or gods in whom, according to his knowledge or superstition, he may believe; but in this case the additional fear of divine punishment often supervenes.

The Strictly Social Virtues at First Alone Regarded—The above view of the origin and nature of the moral sense, which tells us what we ought to do, and of the conscience which reproves us if we disobey it, accords well with what we see of the early and undeveloped condition of this faculty in mankind. The virtues which must be practised, at least generally, by rude men, so that they may associate in a body, are those which are still recognised as the most important. But they are practised almost exclusively in relation to the men of the same tribe; and their opposites are not regarded as crimes in relation to the men of other tribes. No tribe could hold together if murder, robbery, treachery, etc., were common; consequently such crimes within the limits of the same tribe "are branded with everlasting infamy," but excite no such sentiment beyond these limits. A North American Indian is well pleased with himself, and is honoured by others, when he scalps a man of another tribe; and a Dyak cuts off the head of an unoffending person, and dries it as a trophy. The murder of infants has prevailed on the largest scale throughout the world, and has met with no reproach; but infanticide, especially of females, has been thought to be good for the tribe, or at least not injurious. Suicide during former times was not generally considered as a crime, but rather, from the courage displayed, as an honourable act; and it is still practised by some semi-civilised and savage nations without reproach, for it does not obviously concern others of the tribe. It has been recorded that an Indian thug conscientiously regretted that he had not robbed and strangled as many travellers as did his father before him. In a rude state of civilisation the robbery of strangers is, indeed, generally considered as honourable.

Slavery, although in some ways beneficial during ancient times, is a great crime; yet it was not so regarded until quite recently, even by the most civilised nations. And this was especially the case, because the slaves belonged in general to a race different from that of their masters. As barbarians do not regard the opinion of their women, wives are commonly treated like slaves. Most savages are utterly indifferent to the sufferings of strangers, or even delight in witnessing them. It is well known that the women and children of the North American Indians aided in torturing their enemies. Some savages take a horrid pleasure in cruelty to animals, and humanity is an unknown virtue. Nevertheless, besides the family affections, kindness is common, especially during sickness, between the members of the same tribe, and is sometimes extended beyond these limits. Mungo Park's touching account of the kindness of the negro women of the interior to him is well known. Many instances could be given of the noble fidelity of savages towards each other, but not to strangers; common experience justifies the maxim of the Spaniard, "Never, never trust an Indian." There cannot be fidelity without truth; and this fundamental virtue is not rare between the members of the same tribe: thus Mungo Park heard the negro women teaching their young children to love the truth. This, again, is one of the virtues which becomes so deeply rooted in the mind that it is sometimes practised by savages, even at a high cost, towards strangers; but to lie to your enemy has rarely been thought a sin, as the history of modern diplomacy too plainly shows. As soon as a tribe has a recognised leader, disobedience becomes a crime, and even abject submission is looked at as a sacred virtue.

As during rude times no man can be useful or faithful to his tribe without courage, this quality has universally been placed in the highest rank; and although in civilised countries a good yet timid man may be far more useful to the community than a brave one, we cannot help instinctively honouring the latter above a coward, however benevolent. Prudence, on the other hand, which does not concern the welfare of others, though a very useful virtue, has never been highly esteemed. As no man can practise the virtues necessary for the welfare of his tribe without self-sacrifice, self-command, and the power of endurance, these qualities have been at all times highly and most justly valued. The American savage voluntarily submits to the most horrid tortures without a groan, to prove and strengthen his fortitude and courage; and we cannot help admiring him, or even an Indian Fakir, who, from a foolish religious motive, swings suspended by a hook buried in his flesh.

The other so-called self-regarding virtues, which do not obviously, though they may really, affect the welfare of the tribe, have never been esteemed by savages, though now highly appreciated by civilised nations. The greatest intemperance is no reproach with savages. Utter licentiousness, and unnatural crimes, prevail to an astounding extent. As soon, however, as marriage, whether polygamous, or monogamous, becomes common, jealousy will lead to the inculcation of female virtue; and this, being honoured, will tend to spread to the unmarried females. How slowly it spreads to the male sex, we see at the present day. Chastity eminently requires self-command; therefore it has been honoured from a very early period in the moral history of civilised man. As a consequence of this, the senseless practice of celibacy has been ranked from a remote period as a virtue. The hatred of indecency, which appears to us so natural as to be thought innate, and which is so valuable an aid to chastity, is a modern virtue, appertaining exclusively, as Sir G. Staunton remarks, to civilised life. This is shown by the ancient religious rites of various nations, by the drawing on the walls of Pompeii, and by the practices of many savages.

We have now seen that actions are regarded by savages, and were probably so regarded by primeval man, as good or bad, solely as they obviously affect the welfare of the tribe—not that of the species, nor that of an individual member of the tribe. This conclusion agrees well with the belief that the so-called moral sense is aboriginally derived from the social instincts, for both relate at first exclusively to the community.

The chief causes of the low morality of savages, as judged by our standard, are, firstly, the confinement of sympathy to the same tribe. Secondly, powers of reasoning insufficient to recognise the bearing of many virtues, especially of the self-regarding virtues, on the general welfare of the tribe. Savages, for instance, fail to trace the multiplied evils consequent on a want of temperance, chastity, etc. And, thirdly, weak power of self-command; for this power has not been strengthened through long-continued, perhaps inherited, habit, instruction, and religion.

I have entered into the above details on the immorality of savages because some authors have recently taken a high view of their moral nature, or have attributed most of their crimes to mistaken benevolence. These authors appear to rest their conclusion on savages possessing those virtues which are serviceable, or even necessary, for the existence of the family and of the tribe—qualities which they undoubtedly do possess, and often in a high degree.

Concluding Remarks—It was assumed formerly by philosophers of the derivative school of morals that the foundation of morality lay in a form of selfishness; but more recently the "greatest happiness principle" has been brought prominently forward. It is, however, more correct to speak of the latter principle as the standard, and not as the motive of conduct. Nevertheless, all the authors whose works I have consulted, with a few exceptions, write as if there must be a distinct motive for every action, and that this must be associated with some pleasure or displeasure. But man seems often to act impulsively, that is, from instinct or long habit, without any consciousness of pleasure, in the same manner as does probably a bee or ant when it blindly follows its instincts. Under circumstances of extreme peril, as during a fire, when a man endeavours to save a fellow creature without a moment's hesitation, he can hardly feel pleasure; and still less has he time to reflect on the dissatisfaction which he might subsequently experience if he did not make the attempt. Should he afterwards reflect over his own conduct, he would feel that there lies within him an impulsive power widely different from a search after pleasure or happiness; and this seems to be the deeply planted social instinct.

In the case of the lower animals it seems much more appropriate to speak of their social instincts as having been developed for the general good rather than for the general happiness of the species. The term, general good, may be defined as the rearing of the greatest number of individuals in full vigour and health, with all their faculties perfect, under the conditions to which they are subjected. As the social instincts both of man and the lower animals have no doubt been developed by nearly the same steps, it would be advisable, if found practicable, to use the same definition in both cases, and to take as the standard of morality the general good or welfare of the community, rather than the general happiness; but this definition would perhaps require some limitation on account of political ethics.

When a man risks his life to save that of a fellow creature, it seems also more correct to say that he acts for the general good, rather than for the general happiness of mankind. No doubt the welfare and the happiness of the individual usually coincide; and a contented, happy tribe will flourish better than one that is discontented and unhappy. We have seen that even at an early period in the history of man, the expressed wishes of the community will have naturally influenced to a large extent the conduct of each member; and as all wish for happiness, the "greatest happiness principle" will have become a most important secondary guide and object; the social instinct, however, together with sympathy (which leads to our regarding the approbation and disapprobation of others), having served as the primary impulse and guide. Thus the reproach is removed of laying the foundation of the noblest part of our nature in the base principle of selfishness; unless, indeed, the satisfaction which every animal feels, when it follows its proper instincts, and the dissatisfaction felt when prevented, be called selfish.

The wishes and opinions of the members of the same community, expressed at first orally, but later by writing also, either form the sole guides of our conduct, or greatly reinforce the social instincts; such opinions, however, have sometimes a tendency directly opposed to these instincts. This latter fact is well exemplified by the *Law of Honour,* that is, the law of the opinion of our equals, and not of all our countrymen. The breach of this law, even when the breach is known to be strictly accordant with true morality, has caused many a man more agony than a real crime. We recognise the same influence in the burning sense of shame which most of us have felt, even after the interval of years, when calling to mind some accidental breach of a trifling, though fixed, rule of etiquette. The judgment of the community will generally be guided by some rude experience of what is best in the long run for all the members; but this judgment will not rarely err from ignorance and weak powers of reasoning. Hence the strangest customs and superstitions, in complete opposition to the true welfare and happiness of mankind, have become all-powerful throughout the world. We see this in the horror felt by a Hindoo who breaks his caste, and in many other such cases. It would be difficult to distinguish between the remorse felt by a Hindoo who has yielded to the temptation of eating unclean food, from that felt after committing a theft; but the former would probably be the more severe.

How so many absurd rules of conduct, as well as so many absurd religious beliefs, have originated, we do not know; nor how it is that they have become, in all quarters of the world, so deeply impressed on the mind of men; but it is worthy of remark that a belief constantly inculcated during the early years of life, while the brain is impressible, appears to acquire almost the nature of an instinct; and the very essence of an instinct is that it is followed independently of reason. Neither can we say why certain admirable virtues, such as the love of truth, are much more highly appreciated by some savage tribes than by others; nor, again, why similar differences prevail even among highly civilised nations. Knowing how firmly fixed many strange customs and superstitions have become, we need feel no surprise that the self-regarding virtues, supported as they are by reason, should now appear to us so natural as to be thought innate, although they were not valued by man in his early condition.

Notwithstanding many sources of doubt, man can generally and readily distinguish between the higher and lower moral rules. The higher are founded on the social instincts, and relate to the welfare of others. They are supported by the approbation of our fellow men and by reason. The lower rules, though some of them when implying self-sacrifice hardly deserve to be called lower, relate chiefly to self, and arise from public opinion, matured by experience and cultivation; for they are not practised by rude tribes.

As man advances in civilisation, and small tribes are united into larger communities, the simplest reason would tell each individual that he ought to extend his social instincts and sympathies to all the members of the same nation, though personally unknown to him. This point being once reached, there is only an artificial barrier to prevent his sympathies extending to the men of all nations and races. If, indeed, such men are separated from him by great differences in appearance or habits, experience unfortunately shows us how long it is before we look at them as our fellow creatures. Sympathy beyond the confines of man, that is, humanity to the lower animals, seems to be one of the latest moral acquisitions. It is apparently unfelt by savages, except towards their pets. How little the old Romans knew of it is shown by their abhorrent gladiatorial exhibitions. The very idea of humanity, as far as I could observe, was new to most of the Gauchos of the Pampas. This virtue, one of the noblest with which man is endowed, seems to arise, incidentally, from our sympathies becoming more tender and more widely diffused, until they are extended to all sentient beings. As soon as this virtue is honoured and practised by some few men, it spreads through instruction and example to the young, and eventually becomes incorporated in public opinion.

The highest possible stage in moral culture is when we recognise that we ought to control our thoughts, and "not even in inmost thought to think again the sins that made the past so pleasant to us."[4] Whatever makes any bad action familiar to the mind renders its performance by so much the easier. As Marcus Aurelius long ago said, "Such as are thy habitual thoughts, such also will be the character of thy mind; for the soul is dyed by the thoughts."

Our great philosopher, Herbert Spencer, has recently explained his views on the moral sense. He says, "I believe that the experiences of utility organised and consolidated through all past generations of the human race have been producing corresponding modifications, which, by continued transmission and accumulation, have become in us certain faculties of moral intuition—certain emotions responding to right and wrong conduct, which have no apparent basis in the individual experiences of utility." There is not the least inherent improbability, as it seems to me, in virtuous tendencies being more or less strongly inherited; for, not to mention the various dispositions and habits transmitted by many of our domestic animals to their offspring, I have heard of authentic cases in which a desire to steal and

---

[4] Tennyson, "Idylls of the King."

a tendency to lie appeared to run in families of the upper ranks; and as stealing is a rare crime in the wealthy classes, we can hardly account by accidental coincidence for the tendency occurring in two or three members of the same family. If bad tendencies are transmitted, it is probable that good ones are likewise transmitted. That the state of the body by affecting the brain has great influence on the moral tendencies is known to most of those who have suffered from chronic derangements of the digestion or liver. The same fact is likewise shown by the "perversion or destruction of the moral sense being often one of the earliest symptoms of mental derangement"; and insanity is notoriously often inherited. Except through the principle of the transmission of moral tendencies, we cannot understand the differences believed to exist in this respect between the various races of mankind.

Even the partial transmission of virtuous tendencies would be an immense assistance to the primary impulse derived directly and indirectly from the social instincts. Admitting for a moment that virtuous tendencies are inherited, it appears probable, at least in such cases as chastity, temperance, humanity to animals, etc., that they become first impressed on the mental organization through habit, instruction, and example, continued during several generations in the same family, and in a quite subordinate degree, or not at all, by the individuals possessing such virtues having succeeded best in the struggle for life. My chief source of doubt with respect to any such inheritance is that senseless customs, superstitions, and tastes, such as the horror of a Hindoo for unclean food, ought on the same principle to be transmitted. I have not met with any evidence in support of the transmission of superstitious customs or senseless habits, although in itself it is perhaps not less probable than that animals should acquire inherited tastes for certain kinds of food or fear of certain foes.

Finally, the social instincts, which no doubt were acquired by man as by the lower animals for the good of the community, will from the first have given to him some wish to aid his fellows, some feeling of sympathy, and have compelled him to regard their approbation and disapprobation. Such impulses will have served him at a very early period as a rude rule of right and wrong. But as man gradually advanced in intellectual power, and was enabled to trace the more remote consequences of his actions; as he acquired sufficient knowledge to reject baneful customs and superstitions; as he regarded more and more, not only the welfare, but the happiness of his fellow men; as from habit, following on beneficial experience, instruction, and example, his sympathies became more tender and widely diffused, extending to men of all races, to the imbecile, maimed, and other useless members of society, and finally to the lower animals—so would the standard of his morality rise higher and higher. And it is admitted by moralists of the derivative school and by some intuitionists, that the standard of morality has risen since an early period in the history of man.

As a struggle may sometimes be seen going on between the various instincts of the lower animals, it is not surprising that there should be a struggle in man between his social instincts, with their derived virtues, and his lower, though momentarily stronger impulses or desires. This, as Mr. Galton has remarked, is all the less surprising, as man has emerged from a state of barbarism within a comparatively recent period. After having yielded to some temptation we feel a sense of dissatisfaction, shame, repentance, or remorse, analogous to the feelings caused by other powerful instincts or desires, when left unsatisfied or baulked. We compare the weakened impression of a past temptation with the ever-present social instincts, or with habits, gained in early youth and strengthened during our whole lives, until they have become almost as strong as instincts. If with the temptation still before us we do not yield, it is because either the social instinct or some custom is at the moment predominant, or because we have learned that it will appear to us hereafter the stronger, when compared with the weakened impression of the temptation, and we realise that its violation would cause us suffering. Looking to future generations, there is no cause to fear that the social instincts will grow weaker, and we may expect that virtuous habits will grow stronger, becoming perhaps fixed by inheritance. In this case the struggle between our higher and lower impulses will be less severe, and virtue will be triumphant.

WILLIAM SHAKESPEARE's birth date is uncertain, but he was baptized in 1564 in Stratford-on-Avon, England. Many of the details of Shakespeare's life are unknown or contested. His father was a shopkeeper and glover who was a prominent local citizen, serving Stratford as a burgess, an alderman, and as bailiff. William Shakespeare was educated only at the local grammar school, where he studied Latin and most likely read the plays of Terence and Plautus. He married in 1582 and had three children. It is unclear when and how Shakespeare went to London and began working in the theater, but by 1594 he had joined the Lord Chamberlain's Company of players as an actor, a playwright, and a shareholder. Shakespeare's early plays were first performed in the 1590s; his last works were staged in 1613. Shakespeare enjoyed some degree of prosperity, purchasing one of Stratford's largest homes in 1597 and acquiring 107 acres of farmland in 1602. Shakespeare's plays were published in his lifetime, but not with his authorization. He died in 1616.

From *The Tragedy of Othello, The Moor of Venice,* edited by Lawrence Mason. Publisher: Yale University Press, 1918.

Othello

CHARACTERS

OTHELLO, the Moor
BRABANTIO, father to Desdemona
CASSIO, an honourable Lieutenant
IAGO, a villain [ancient to Othello]
RODERIGO, a gulled gentleman
DUKE OF VENICE
SENATORS
MONTANO, Governor of Cyprus [before Othello]
GENTLEMEN of Cyprus
LODOVICO and GRATIANO
two noble Venetians [kinsman and brother, respectively, to Brabantio]
SAILORS
CLOWN [in Othello's retinue]
DESDEMONA, wife to Othello
[and daughter to Brabantio]
EMILIA, wife to Iago
[and maid-in-waiting to Desdemona]
BIANCA, a courtesan [in love with Cassio]

MESSENGER, HERALD, OFFICERS, MUSICIANS, and ATTENDANTS

SCENE: *Act I, at Venice; Acts II-V, at a sea-port in Cyprus.*

ACT I

SCENE ONE—*Venice. A Street.*

Enter RODERIGO *and* IAGO.

RODERIGO: Tush! Never tell me; I take it much unkindly
That thou, Iago, who hast had my purse
As if the strings were thine, shouldst know of this.
IAGO: 'Sblood, but you will not hear me:
If ever I did dream of such a matter,
Abhor me.
RODERIGO: Thou told'st me thou didst hold him in thy hate.
IAGO: Despise me if I do not. Three great ones of the city,
In personal suit to make me his lieutenant,
Off-capp'd to him; and, by the faith of man,
I know my price, I am worth no worse a place;
But he, as loving his own pride and purposes,
Evades them, with a bombast circumstance
Horribly stuff'd with epithets of war;
And, in conclusion,
Nonsuits[1] my mediators; for, "Certes," says he,
"I have already chose my officer."
And what was he?
Forsooth, a great arithmetician,
One Michael Cassio, a Florentine,
A fellow almost damn'd in a fair wife;
That never set a squadron in the field,
Nor the division of a battle knows
More than a spinster; unless the bookish theoric,
Wherein the toged consuls can propose
As masterly as he: mere prattle, without practice,
Is all his soldiership. But he, sir, had the election;
And I—of whom his eyes had seen the proof
At Rhodes, at Cyprus, and on other grounds

---

[1] [*Nonsuits:* rebuffs.]

Christian and heathen—must be be-lee'd and calm'd
By debitor and creditor; this counter-caster,[2]
He, in good time, must his lieutenant be,
And I—God bless the mark!—his Moorship's ancient.[3]

RODERIGO: By heaven, I rather would have been his hangman.

IAGO: Why, there's no remedy: 'tis the curse of service,
Preferment goes by letter and affection,
Not by the old gradation, where each second
Stood heir to the first. Now, sir, be judge yourself,
Whether I in any just term am affin'd
To love the Moor.

RODERIGO: I would not follow him then.

IAGO: O! sir, content you;
I follow him to serve my turn upon him;
We cannot all be masters, nor all masters
Cannot be truly follow'd. You shall mark
Many a duteous and knee-crooking knave,
That, doting on his own obsequious bondage,
Wears out his time, much like his master's ass,
For nought but provender, and when he's old, cashier'd;
Whip me such honest knaves. Others there are
Who, trimm'd in forms and visages of duty,
Keep yet their hearts attending on themselves,
And, throwing but shows of service on their lords,
Do well thrive by them, and when they have lin'd their coats
Do themselves homage: these fellows have some soul;
And such a one do I profess myself. For, sir,
It is as sure as you are Roderigo,
Were I the Moor, I would not be Iago:
In following him, I follow but myself;
Heaven is my judge, not I for love and duty,
But seeming so, for my peculiar end:
For when my outward action doth demonstrate

---

[2] [*counter-caster:* reckoner, caster-up of accounts.]

[3] [*ancient:* ensign.]

The native act and figure of my heart
In complement extern, 'tis not long after
But I will wear my heart upon my sleeve
For daws to peck at: I am not what I am.
RODERIGO: What a full fortune does the thick-lips owe,[4]
If he can carry 't thus!
IAGO: Call up her father;
Rouse him, make after him, poison his delight,
Proclaim him in the streets, incense her kinsmen,
And, though he in a fertile climate dwell,
Plague him with flies; though that his joy be joy,
Yet throw such chances of vexation on 't
As it may lose some colour.
RODERIGO: Here is her father's house; I'll call aloud.
IAGO: Do; with like timorous accent and dire yell
As when, by night and negligence, the fire
Is spied in populous cities.
RODERIGO: What, ho! Brabantio! Signior Brabantio, ho!
IAGO: Awake! what, ho! Brabantio! thieves! thieves! thieves!
Look to your house, your daughter, and your bags!
Thieves! thieves!

BRABANTIO [*appears*] *above at a window.*

BRABANTIO: What is the reason of this terrible summons?
What is the matter there?
RODERIGO: Signior, is all your family within?
IAGO: Are your doors lock'd?
BRABANTIO: Why? Wherefore ask you this?
IAGO: 'Zounds! sir, you're robb'd; for shame, put on your gown;
Your heart is burst, you have lost half your soul;
Even now, now, very now, an old black ram
Is tupping your white ewe. Arise, arise!
Awake the snorting citizens with the bell,

---

[4] [*owe:* own.]

Or else the devil will make a grandsire of you.
Arise, I say.

BRABANTIO: What! have you lost your wits?

RODERIGO: Most reverend signior, do you know my voice?

BRABANTIO: Not I, what are you?

RODERIGO: My name is Roderigo.

BRABANTIO: The worser welcome:
I have charg'd thee not to haunt about my doors:
In honest plainness thou hast heard me say
My daughter is not for thee; and now, in madness,
Being full of supper and distempering draughts,
Upon malicious knavery dost thou come
To start my quiet.

RODERIGO: Sir, sir, sir!

BRABANTIO: But thou must needs be sure
My spirit and my place have in them power
To make this bitter to thee.

RODERIGO: Patience, good sir.

BRABANTIO: What tell'st thou me of robbing? this is Venice;
My house is not a grange.

RODERIGO: Most grave Brabantio,
In simple and pure soul I come to you.

IAGO: 'Zounds! sir, you are one of those that will not serve God if the devil bid you. Because we come to do you service and you think we are ruffians, you'll have your daughter covered with a Barbary horse; you'll have your nephews neigh to you; you'll have coursers for cousins and gennets for germans.[5]

BRABANTIO: What profane wretch art thou?

IAGO: I am one, sir, that comes to tell you, your daughter and the Moor are now making the beast with two backs.

BRABANTIO: Thou art a villain.

IAGO: You are—a senator.

---

[5] [*gennets:* small Spanish horses; *germans:* near relatives.]

BRABANTIO: This thou shalt answer; I know thee, Roderigo.
RODERIGO: Sir, I will answer any thing. But, I beseech you,
If 't be your pleasure and most wise consent,—
As partly, I find, it is,—that your fair daughter,
At this odd-even and dull watch o' the night,
Transported with no worse nor better guard
But with a knave of common hire, a gondolier,
To the gross clasps of a lascivious Moor,—
If this be known to you, and your allowance,
We then have done you bold and saucy wrongs;
But if you know not this, my manners tell me
We have your wrong rebuke. Do not believe
That, from the sense of all civility,
I thus would play and trifle with your reverence:
Your daughter, if you have not given her leave,
I say again, hath made a gross revolt;
Tying her duty, beauty, wit and fortunes
In an extravagant and wheeling stranger
Of here and everywhere. Straight satisfy yourself:
If she be in her chamber or your house,
Let loose on me the justice of the state
For thus deluding you.
BRABANTIO: Strike on the tinder, ho!
Give me a taper! call up all my people!
This accident is not unlike my dream;
Belief of it oppresses me already.
Light, I say! light! *Exit.*
IAGO: Farewell, for I must leave you:
It seems not meet nor wholesome to my place
To be produc'd, as, if I stay, I shall,
Against the Moor; for, I do know the state,
However this may gall him with some check,
Cannot with safety cast him; for he's embark'd
With such loud reason to the Cyprus wars,—
Which even now stand in act—that, for their souls,
Another of his fathom they have none,

To lead their business; in which regard,
Though I do hate him as I do hell-pains,
Yet, for necessity of present life,
I must show out a flag and sign of love,
Which is indeed but sign. That you shall surely find him,
Lead to the Sagittary the raised search;
And there will I be with him. So, farewell. *Exit.*

Enter [below] BRABANTIO, *with* Servants *and torches.*

BRABANTIO: It is too true an evil: gone she is,
And what's to come of my despised time
Is nought but bitterness. Now, Roderigo,
Where didst thou see her? O, unhappy girl!
With the Moor, sayst thou? Who would be a father!
How didst thou know 'twas she? O, she deceives me
Past thought. What said she to you? Get more tapers!
Raise all my kindred! Are they married, think you?
RODERIGO: Truly, I think they are.
BRABANTIO: O heaven! How got she out? O, treason of the blood:
Fathers, from hence trust not your daughters' minds
By what you see them act. Are there not charms
By which the property of youth and maidhood
May be abus'd? Have you not read, Roderigo,
Of some such thing?
RODERIGO: Yes, sir, I have indeed.
BRABANTIO: Call up my brother. O! that you had had her.
Some one way, some another! Do you know
Where we may apprehend her and the Moor?
RODERIGO: I think I can discover him, if you please
To get good guard and go along with me.
BRABANTIO: Pray you, lead on. At every house I'll call;
I may command at most. Get weapons, ho!
And raise some special officers of night.
On, good Roderigo; I'll deserve your pains. *Exeunt.*

SCENE TWO—*Another Street.*

Enter OTHELLO, IAGO, *and* Attendants, *with torches.*

IAGO: Though in the trade of war I have slain men,
Yet do I hold it very stuff o' the conscience
To do no contriv'd murder: I lack iniquity
Sometimes to do me service. Nine or ten times
I had thought to have yerk'd him here under the ribs.
OTHELLO: 'Tis better as it is.
IAGO: Nay, but he prated,
And spoke such scurvy and provoking terms
Against your honour
That, with the little godliness I have,
I did full hard forbear him. But, I pray, sir,
Are you fast married? Be assur'd of this,
That the magnifico is much belov'd,
And hath in his effect a voice potential
As double as the duke's; he will divorce you,
Or put upon you what restraint and grievance
The law—with all his might to enforce it on—
Will give him cable.
OTHELLO: Let him do his spite:
My services which I have done the signiory
Shall out-tongue his complaints. 'Tis yet to know,
Which when I know that boasting is an honour
I shall promulgate, I fetch my life and being
From men of royal siege, and my demerits
May speak unbonneted[6] to as proud a fortune
As this that I have reach'd; for know, Iago,
But that I love the gentle Desdemona,
I would not my unhoused free condition
Put into circumscription and confine
For the sea's worth. But, look! what lights come yond?

---

6 [*unbonneted:* without saluting, on equal terms.]

IAGO: Those are the raised father and his friends:
You were best go in.
OTHELLO: Not I; I must be found:
My parts, my title, and my perfect soul
Shall manifest me rightly. Is it they?
IAGO: By Janus, I think no.

Enter CASSIO, *with* Officers, *and torches.*

OTHELLO: The servants of the duke, and my lieutenant.
The goodness of the night upon you, friends!
What is the news?
CASSIO: The duke does greet you, general,
And he requires your haste-post-haste appearance,
Even on the instant.
OTHELLO: What is the matter, think you?
CASSIO: Something from Cyprus, as I may divine.
It is a business of some heat; the galleys
Have sent a dozen sequent messengers
This very night at one another's heels,
And many of the consuls, rais'd and met,
Are at the duke's already. You have been hotly call'd for;
When, being not at your lodging to be found,
The senate hath sent about three several quests
To search you out.
OTHELLO: 'Tis well I am found by you.
I will but spend a word here in the house,
And go with you. [*Exit.*]
CASSIO: Ancient, what makes he here?
IAGO: Faith, he to-night hath boarded a land carrack;
If it prove lawful prize, he's made for ever.
CASSIO: I do not understand.
IAGO: He's married.
CASSIO: To who?
IAGO: Marry, to—

[*Enter* OTHELLO.]

Come, captain, will you go?
OTHELLO: Have with you.
CASSIO: Here comes another troop to seek for you.
IAGO: It is Brabantio. General, be advis'd;
He comes to bad intent.

Enter BRABANTIO, RODERIGO, *with* Officers, *and torches.*

OTHELLO: Holla! stand there!
RODERIGO: Signior, it is the Moor.
BRABANTIO: Down with him, thief!
[*They draw on both sides.*]

IAGO: You, Roderigo! come, sir, I am for you.
OTHELLO: Keep up your bright swords, for the dew will rust them.
Good signior, you shall more command with years
Than with your weapons.
BRABANTIO: O thou foul thief! where hast thou stow'd my daughter?
Damn'd as thou art, thou hast enchanted her;
For I'll refer me to all things of sense,
If she in chains of magic were not bound,
Whether a maid so tender, fair, and happy,
So opposite to marriage that she shunn'd
The wealthy curled darlings of our nation,
Would ever have, to incur a general mock,
Run from her guardage to the sooty bosom
Of such a thing as thou; to fear, not to delight.
Judge me the world, if 'tis not gross in sense
That thou hast practis'd on her with foul charms,
Abus'd her delicate youth with drugs or minerals
That weaken motion: I'll have 't disputed on;
'Tis probable, and palpable to thinking.
I therefore apprehend and do attach thee
For an abuser of the world, a practiser
Of arts inhibited and out of warrant.

Lay hold upon him: if he do resist,
Subdue him at his peril.

OTHELLO: Hold your hands,
Both you of my inclining, and the rest:
Were it my cue to fight, I should have known it
Without a prompter. Where will you that I go
To answer this your charge?

BRABANTIO: To prison; till fit time
Of law and course of direct session
Call thee to answer.

OTHELLO: What if I do obey?
How may the duke be therewith satisfied,
Whose messengers are here about my side,
Upon some present business of the state
To bring me to him?

OFFICER: 'Tis true, most worthy signior;
The duke's in council, and your noble self,
I am sure, is sent for.

BRABANTIO: How! the duke in council!
In this time of the night! Bring him away.
Mine's not an idle cause: the duke himself,
Or any of my brothers of the state,
Cannot but feel this wrong as 'twere their own;
For if such actions may have passage free,
Bond-slaves and pagans shall our statesmen be. *Exeunt.*

SCENE THREE—*A Council Chamber.*

Enter DUKE, *and* SENATORS *set at a table, with lights and* Attendants.

DUKE: There is no composition in these news
That gives them credit.

FIRST SENATOR: Indeed, they are disproportion'd;
My letters say a hundred and seven galleys.

DUKE: And mine, a hundred forty.
SECOND SENATOR: And mine, two hundred:
But though they jump not on a just account,—
As in these cases, where the aim reports,
'Tis oft with difference,—yet do they all confirm
A Turkish fleet, and bearing up to Cyprus.
DUKE: Nay, it is possible enough to judgment:
I do not so secure me in the error,
But the main article I do approve
In fearful sense.
SAILOR [*within*]: What, ho! what, ho! what, ho!
OFFICER: A messenger from the galleys.

Enter SAILOR.

DUKE: Now, what's the business?
SAILOR: The Turkish preparation makes for Rhodes;
So was I bid report here to the state
By Signior Angelo.
DUKE: How say you by this change?
FIRST SENATOR: This cannot be,
By no assay of reason; 'tis a pageant
To keep us in false gaze. When we consider
The importancy of Cyprus to the Turk,
And let ourselves again but understand,
That as it more concerns the Turk than Rhodes,
So may he with more facile question bear it,
For that it stands not in such warlike brace,
But altogether lacks the abilities
That Rhodes is dress'd in: if we make thought of this,
We must not think the Turk is so unskilful
To leave that latest which concerns him first,
Neglecting an attempt of ease and gain,
To wake and wage a danger profitless.
DUKE: Nay, in all confidence, he's not for Rhodes.

OFFICER: Here is more news.

Enter a MESSENGER.

MESSENGER: The Ottomites, reverend and gracious,
Steering with due course toward the isle of Rhodes,
Have there injointed them with an after fleet.
FIRST SENATOR: Ay, so I thought. How many, as you guess?
MESSENGER: Of thirty sail; and now they do re-stem
Their backward course, bearing with frank appearance
Their purposes toward Cyprus. Signior Montano,
Your trusty and most valiant servitor,
With his free duty recommends you thus,
And prays you to believe him.
DUKE: 'Tis certain then, for Cyprus.
Marcus Luccicos, is not he in town?
FIRST SENATOR: He's now in Florence.
DUKE: Write from us to him; post-post-haste dispatch.
FIRST SENATOR: Here comes Brabantio and the valiant Moor.

Enter BRABANTIO, OTHELLO, CASSIO, IAGO, RODERIGO, *and* Officers.

DUKE: Valiant Othello, we must straight employ you
Against the general enemy Ottoman.
[*To* BRABANTIO.] I did not see you; welcome, gentle signior;
We lack'd your counsel and your help to-night.
BRABANTIO: So did I yours. Good your grace, pardon me;
Neither my place nor aught I heard of business
Hath rais'd me from my bed, nor doth the general care
Take hold of me, for my particular grief
Is of so flood-gate and o'erbearing nature
That it engluts and swallows other sorrows
And it is still itself.
DUKE: Why, what's the matter?
BRABANTIO: My daughter! O! my daughter.

SENATOR: Dead?
BRABANTIO: Ay, to me;
She is abus'd, stol'n from me, and corrupted
By spells and medicines bought of mountebanks;
For nature so preposterously to err,
Being not deficient, blind, or lame of sense,
Sans witchcraft could not.
DUKE: Whoe'er he be that in this foul proceeding
Hath thus beguil'd your daughter of herself
And you of her, the bloody book of law
You shall yourself read in the bitter letter
After your own sense; yea, though our proper son
Stood in your action.
BRABANTIO: Humbly I thank your Grace.
Here is the man, this Moor; whom now, it seems,
Your special mandate for the state affairs,
Hath hither brought.
ALL: We are very sorry for it.
DUKE [*to* OTHELLO]: What, in your own part, can you say to this?
BRABANTIO: Nothing, but this is so.
OTHELLO: Most potent, grave, and reverend signiors,
My very noble and approv'd good masters,
That I have ta'en away this old man's daughter,
It is most true; true, I have married her:
The very head and front of my offending
Hath this extent, no more. Rude am I in my speech,
And little bless'd with the soft phrase of peace;
For since these arms of mine had seven years' pith,
Till now some nine moons wasted, they have us'd
Their dearest action in the tented field;
And little of this great world can I speak,
More than pertains to feats of broil and battle;
And therefore little shall I grace my cause
In speaking for myself. Yet, by your gracious patience,

I will a round unvarnish'd tale deliver
Of my whole course of love; what drugs, what charms,
What conjuration, and what mighty magic,
For such proceeding I am charg'd withal,
I won his daughter.

BRABANTIO: A maiden never bold;
Of spirit so still and quiet, that her motion
Blush'd at herself; and she, in spite of nature,
Of years, of country, credit, everything,
To fall in love with what she fear'd to look on!
It is a judgment maim'd and most imperfect
That will confess perfection so could err
Against all rules of nature, and must be driven
To find out practices of cunning hell,
Why this should be. I therefore vouch again
That with some mixtures powerful o'er the blood,
Or with some dram conjur'd to this effect,
He wrought upon her.

DUKE: To vouch this, is no proof,
Without more certain and more overt test
Than these thin habits and poor likelihoods
Of modern seeming do prefer against him.

FIRST SENATOR: But, Othello, speak:
Did you by indirect and forced courses
Subdue and poison this young maid's affections;
Or came it by request and such fair question
As soul to soul affordeth?

OTHELLO: I do beseech you,
Send for the lady to the Sagittary,
And let her speak of me before her father:
If you do find me foul in her report,
The trust, the office I do hold of you,
Not only take away, but let your sentence
Even fall upon my life.

DUKE: Fetch Desdemona hither.

OTHELLO: Ancient, conduct them; you best know the place.
[*Exeunt* IAGO *and* Attendants.]
And, till she come, as truly as to heaven
I do confess the vices of my blood,
So justly to your grave ears I'll present
How I did thrive in this fair lady's love,
And she in mine.
DUKE: Say it, Othello.
OTHELLO: Her father lov'd me; oft invited me;
Still question'd me the story of my life
From year to year, the battles, sieges, fortunes
That I have pass'd.
I ran it through, even from my boyish days
To the very moment that he bade me tell it;
Wherein I spake of most disastrous chances,
Of moving accidents by flood and field,
Of hair-breadth 'scapes i' the imminent deadly breach,
Of being taken by the insolent foe
And sold to slavery, of my redemption thence
And portance in my traveller's history;
Wherein of antres vast and deserts idle,
Rough quarries, rocks and hills whose heads touch heaven,
It was my hint to speak, such was the process;
And of the Cannibals that each other eat,
The Anthropophagi, and men whose heads
Do grow beneath their shoulders. This to hear
Would Desdemona seriously incline;
But still the house-affairs would draw her thence;
Which ever as she could with haste dispatch,
She'd come again, and with a greedy ear
Devour up my discourse. Which I observing,
Took once a pliant hour, and found good means
To draw from her a prayer of earnest heart
That I would all my pilgrimage dilate,
Whereof by parcels she had something heard,

But not intentively: I did consent;
And often did beguile her of her tears,
When I did speak of some distressful stroke
That my youth suffer'd. My story being done,
She gave me for my pains a world of sighs:
She swore, in faith, 'twas strange, 'twas passing strange;
'Twas pitiful, 'twas wondrous pitiful:
She wish'd she had not heard it, yet she wish'd
That heaven had made her such a man; she thank'd me,
And bade me, if I had a friend that lov'd her,
I should but teach him how to tell my story,
And that would woo her. Upon this hint I spake:
She lov'd me for the dangers I had pass'd,
And I lov'd her that she did pity them.
This only is the witchcraft I have us'd:
Here comes the lady; let her witness it.

Enter DESDEMONA, IAGO, *and* Attendants.

DUKE: I think this tale would win my daughter too.
Good Brabantio,
Take up this mangled matter at the best;
Men do their broken weapons rather use
Than their bare hands.
BRABANTIO: I pray you, hear her speak:
If she confess that she was half the wooer,
Destruction on my head, if my bad blame
Light on the man! Come hither, gentle mistress:
Do you perceive in all this noble company
Where most you owe obedience?
DESDEMONA: My noble father,
I do perceive here a divided duty:
To you I am bound for life and education;
My life and education both do learn me
How to respect you; you are the lord of duty,
I am hitherto your daughter: but here's my husband;
And so much duty as my mother show'd

To you, preferring you before her father,
So much I challenge that I may profess
Due to the Moor my lord.
BRABANTIO: God be with you! I have done.
Please it your Grace, on to the state affairs:
I had rather to adopt a child than get it.
Come hither, Moor:
I here do give thee that with all my heart
Which, but thou hast already, with all my heart
I would keep from thee. For your sake, jewel,
I am glad at soul I have no other child;
For thy escape would teach me tyranny,
To hang clogs on them. I have done, my lord.
DUKE: Let me speak like yourself and lay a sentence,
Which as a grise or step, may help these lovers
Into your favour.
When remedies are past, the griefs are ended
By seeing the worst, which late on hopes depended.
To mourn a mischief that is past and gone
Is the next way to draw new mischief on.
What cannot be preserv'd when Fortune takes,
Patience her injury a mockery makes.
The robb'd that smiles steals something from the thief;
He robs himself that spends a bootless grief.
BRABANTIO: So let the Turk of Cyprus us beguile;
We lose it not so long as we can smile.
He bears the sentence well that nothing bears
But the free comfort which from thence he hears;
But he bears both the sentence and the sorrow
That, to pay grief, must of poor patience borrow.
These sentences, to sugar, or to gall,
Being strong on both sides, are equivocal:
But words are words; I never yet did hear
That the bruis'd heart was pierced through the ear.
I humbly beseech you, proceed to the affairs of state.

DUKE: The Turk with a most mighty preparation makes for Cyprus. Othello, the fortitude of the place is best known to you; and though we have there a substitute of most allowed sufficiency, yet opinion, a sovereign mistress of effects, throws a more safer voice on you: you must therefore be content to slubber the gloss of your new fortunes with this more stubborn and boisterous expedition.

OTHELLO: The tyrant custom, most grave senators,
Hath made the flinty and steel couch of war
My thrice-driven bed of down: I do agnize
A natural and prompt alacrity
I find in hardness, and do undertake
These present wars against the Ottomites.
Most humbly therefore bending to your state,
I crave fit disposition for my wife,
Due reference of place and exhibition,
With such accommodation and besort
As levels with her breeding.

DUKE: If you please,
Be 't at her father's.

BRABANTIO: I'll not have it so.

OTHELLO: Nor I.

DESDEMONA: Nor I; I would not there reside,
To put my father in impatient thoughts
By being in his eye. Most gracious duke,
To my unfolding lend your gracious ear;
And let me find a charter in your voice
To assist my simpleness.

DUKE: What would you, Desdemona?

DESDEMONA: That I did love the Moor to live with him,
My downright violence and storm of fortunes
May trumpet to the world; my heart's subdu'd
Even to the very quality of my lord;
I saw Othello's visage in his mind,
And to his honours and his valiant parts
Did I my soul and fortunes consecrate.

So that, dear lords, if I be left behind,
A moth of peace, and he go to the war,
The rites for which I love him are bereft me,
And I a heavy interim shall support
By his dear absence. Let me go with him.
OTHELLO: Let her have your voices.
Vouch with me, heaven, I therefore beg it not
To please the palate of my appetite,
Nor to comply with heat,—the young affects[7]
In me defunct,—and proper satisfaction,
But to be free and bounteous to her mind;
And heaven defend your good souls that you think
I will your serious and great business scant
For she is with me. No, when light-wing'd toys
Of feather'd Cupid seel with wanton dulness
My speculative and offic'd instruments,
That my disports corrupt and taint my business,
Let housewives make a skillet of my helm,
And all indign and base adversities
Make head against my estimation!
DUKE: Be it as you shall privately determine,
Either for her stay or going. The affair cries haste,
And speed must answer it.
SENATOR: You must away to-night.
OTHELLO: With all my heart.
DUKE: At nine i' the morning here we'll meet again.
Othello, leave some officer behind,
And he shall our commission bring to you;
With such things else of quality and respect
As doth import you.
OTHELLO: So please your Grace, my ancient;
A man he is of honesty and trust:
To his conveyance I assign my wife,
With what else needful your good grace shall think

[7] [*young affects:* passions of youth.]

To be sent after me.

DUKE: Let it be so.
Good night to every one. [*To* BRABANTIO.] And, noble signior,
If virtue no delighted beauty lack,
Your son-in-law is far more fair than black.

SENATOR: Adieu, brave Moor! use Desdemona well.

BRABANTIO: Look to her, Moor, if thou hast eyes to see:
She has deceiv'd her father, and may thee.

Exit [*with* DUKE, SENATORS, Officers, *etc.*].

OTHELLO: My life upon her faith! Honest Iago,
My Desdemona must I leave to thee:
I prithee, let thy wife attend on her;
And bring them after in the best advantage.
Come, Desdemona; I have but an hour
Of love, of worldly matters and direction,
To spend with thee: we must obey the time.

Exit [*with* DESDEMONA].

RODERIGO: Iago!

IAGO: What sayst thou, noble heart?

RODERIGO: What will I do, think'st thou?

IAGO: Why, go to bed, and sleep.

RODERIGO: I will incontinently drown myself.

IAGO: Well, if thou dost, I shall never love thee after. Why, thou silly gentleman!

RODERIGO: It is silliness to live when to live is torment; and then have we a prescription to die when death is our physician.

IAGO: O! villainous; I have looked upon the world for four times seven years, and since I could distinguish betwixt a benefit and an injury, I never found man that knew how to love himself. Ere I would say, I would drown myself for the love of a guinea-hen, I would change my humanity with a baboon.

RODERIGO: What should I do? I confess it is my shame to be so fond; but it is not in my virtue to amend it.

IAGO: Virtue! a fig! 'tis in ourselves that we are thus, or thus. Our bodies are our gardens, to the which our wills are gardeners; so that if we will plant nettles or sow lettuce, set hyssop and weed up thyme, supply it with one gender of herbs or distract it with many, either to have it sterile with idleness or manured with industry, why, the power and corrigible authority of this lies in our wills. If the balance of our lives had not one scale of reason to poise another of sensuality, the blood and baseness of our natures would conduct us to most preposterous conclusions; but we have reason to cool our raging motions, our carnal stings, our unbitted lusts, whereof I take this that you call love to be a sect or scion.

RODERIGO: It cannot be.

IAGO: It is merely a lust of the blood and a permission of the will. Come, be a man. Drown thyself! drown cats and blind puppies. I have professed me thy friend, and I confess me knit to thy deserving with cables of perdurable toughness; I could never better stead thee than now. Put money in thy purse; follow these wars; defeat thy favour with a usurped beard; I say, put money in thy purse. It cannot be that Desdemona should long continue her love to the Moor—put money in thy purse—nor he his to her. It was a violent commencement in her, and thou shalt see an answerable sequestration; put but money in thy purse. These Moors are changeable in their wills—fill thy purse with money. The food that to him now is as luscious as locusts, shall be to him shortly as bitter as coloquintida. She must change for youth: when she is sated with his body, she will find the error of her choice. She must have change, she must: therefore put money in thy purse. If thou wilt needs damn thyself, do it a more delicate way than drowning. Make all the money thou canst. If sanctimony and a frail vow betwixt an erring barbarian and a supersubtle

Venetian be not too hard for my wits and all the tribe of hell, thou shalt enjoy her; therefore make money. A pox of drowning thyself! it is clean out of the way: seek thou rather to be hanged in compassing thy joy than to be drowned and go without her.

RODERIGO: Wilt thou be fast to my hopes, if I depend on the issue?

IAGO: Thou art sure of me: go, make money. I have told thee often, and I re-tell thee again and again, I hate the Moor: my cause is hearted: thine hath no less reason. Let us be conjunctive in our revenge against him; if thou canst cuckold him, thou dost thyself a pleasure, me a sport. There are many events in the womb of time which will be delivered. Traverse; go: provide thy money. We will have more of this to-morrow. Adieu.

RODERIGO: Where shall we meet i' the morning?

IAGO: At my lodging.

RODERIGO: I'll be with thee betimes.

IAGO: Go to; farewell. Do you hear, Roderigo?

RODERIGO: What say you?

IAGO: No more of drowning, do you hear?

RODERIGO: I am chang'd. I'll go sell all my land. *Exit.*

IAGO: Thus do I ever make my fool my purse;
For I mine own gain'd knowledge should profane,
If I would time expend with such a snipe
But for my sport and profit. I hate the Moor,
And it is thought abroad that 'twixt my sheets
He has done my office: I know not if 't be true,
But I, for mere suspicion in that kind,
Will do as if for surety. He holds me well;
The better shall my purpose work on him.
Cassio's a proper man; let me see now:
To get his place; and to plume up my will
In double knavery; how, how? Let's see:
After some time to abuse Othello's ear
That he is too familiar with his wife:

He hath a person and a smooth dispose
To be suspected; fram'd to make women false.
The Moor is of a free and open nature,
That thinks men honest that but seem to be so,
And will as tenderly be led by the nose
As asses are.
I have 't; it is engender'd: hell and night
Must bring this monstrous birth to the world's light.

[*Exit.*]

ACT II

SCENE ONE—*A sea-port town in Cyprus. An open place near the Quay.*

Enter MONTANO *and two* GENTLEMEN.

MONTANO: What from the cape can you discern at sea?
FIRST GENTLEMAN: Nothing at all: it is a high-wrought flood;
I cannot 'twixt the heaven and the main
Descry a sail.
MONTANO: Methinks the wind hath spoke aloud at land;
A fuller blast ne'er shook our battlements;
If it hath ruffian'd so upon the sea,
What ribs of oak, when mountains melt on them,
Can hold the mortise?[8] what shall we hear of this?
SECOND GENTLEMAN: A segregation of the Turkish fleet;
For do but stand upon the foaming shore,
The chidden billow seems to pelt the clouds;
The wind-shak'd surge, with high and monstrous mane,
Seems to cast water on the burning bear
And quench the guards of the ever-fixed pole:
I never did like molestation view
On the enchafed flood.
MONTANO: If that the Turkish fleet
Be not enshelter'd and embay'd, they are drown'd;
It is impossible they bear it out.

Enter a [THIRD] GENTLEMAN.

THIRD GENTLEMAN: News, lads! our wars are done.
The desperate tempest hath so bang'd the Turks
That their designment halts; a noble ship of Venice
Hath seen a grievous wrack and sufferance
On most part of their fleet.

---

[8] [*hold the mortise:* remain unshattered.]

MONTANO: How! is this true?
THIRD GENTLEMAN: The ship is here put in,
A Veronesa; Michael Cassio,
Lieutenant to the warlike Moor Othello,
Is come on shore: the Moor himself's at sea,
And is in full commission here for Cyprus.
MONTANO: I am glad on 't; 'tis a worthy governor.
THIRD GENTLEMAN: But this same Cassio, though he speak of comfort
Touching the Turkish loss, yet he looks sadly
And prays the Moor be safe; for they were parted
With foul and violent tempest.
MONTANO: Pray heaven he be;
For I have serv'd him, and the man commands
Like a full soldier. Let's to the sea-side, ho!
As well to see the vessel that's come in
As to throw out our eyes for brave Othello,
Even till we make the main and the aerial blue
An indistinct regard.
THIRD GENTLEMAN: Come, let's do so;
For every minute is expectancy
Of more arrivance.

Enter CASSIO.

CASSIO: Thanks, you the valiant of this warlike isle,
That so approve the Moor. O! let the heavens
Give him defence against the elements,
For I have lost him on a dangerous sea.
MONTANO: Is he well shipp'd?
CASSIO: His bark is stoutly timber'd, and his pilot
Of very expert and approv'd allowance;
Therefore my hopes, not surfeited to death,
Stand in bold cure.
[*Cry*] *within.* "A sail!—a sail!—a sail!"

Enter a MESSENGER.

CASSIO: What noise?
MESSENGER: The town is empty; on the brow o' the sea
  Stand ranks of people, and they cry, "A sail!"
CASSIO: My hopes do shape him for the governor.
[*Guns heard.*]

MESSENGER: They do discharge their shot of courtesy;
  Our friends, at least.
CASSIO: I pray you, sir, go forth,
  And give us truth who 'tis that is arriv'd.
MESSENGER: I shall. *Exit.*
MONTANO: But, good lieutenant, is your general wiv'd?
CASSIO: Most fortunately: he hath achiev'd a maid
  That paragons description and wild fame;
  One that excels the quirks of blazoning pens,
  And in th' essential vesture of creation
  Does tire the ingener.[9]

Enter [SECOND] GENTLEMAN.

How now! who has put in?
SECOND GENTLEMAN: 'Tis one Iago, ancient to the general.
CASSIO: He has had most favourable and happy speed:
  Tempests themselves, high seas, and howling winds,
  The gutter'd rocks, and congregated sands,
  Traitors ensteep'd to clog the guiltless keel,
  As having sense of beauty, do omit
  Their mortal natures, letting go safely by
  The divine Desdemona.
MONTANO: What is she?
CASSIO: She that I spake of, our great captain's captain,
  Left in the conduct of the bold Iago,
  Whose footing here anticipates our thoughts
  A se'nnight's speed. Great Jove, Othello guard,
  And swell his sail with thine own powerful breath,

---

9 [*ingener:* eulogist.]

That he may bless this bay with his tall ship,
Make love's quick pants in Desdemona's arms,
Give renew'd fire to our extincted spirits,
And bring all Cyprus comfort!

Enter DESDEMONA, IAGO, RODERIGO, *and* EMILIA [*with* Attendants].

O! behold,
The riches of the ship is come on shore.
Ye men of Cyprus, let her have your knees.
Hail to thee, lady! and the grace of heaven,
Before, behind thee, and on every hand,
Enwheel thee round!

DESDEMONA: I thank you, valiant Cassio.
What tidings can you tell me of my lord?

CASSIO: He is not yet arriv'd; nor know I aught
But that he's well, and will be shortly here.

DESDEMONA: O! but I fear—How lost you company?

CASSIO: The great contention of the sea and skies
Parted our fellowship. But hark! a sail.
[*Cry*] *within.* "A sail!—a sail!" [*Guns heard.*]

SECOND GENTLEMAN: They give their greeting to the citadel:
This likewise is a friend.

CASSIO: See for the news!
[*Exit* SECOND GENTLEMAN.]
Good ancient, you are welcome. [*To* EMILIA.] Welcome, mistress:
Let it not gall your patience, good Iago,
That I extend my manners; 'tis my breeding
That gives me this bold show of courtesy. [*Kissing her.*]

IAGO: Sir, would she give you so much of her lips
As of her tongue she oft bestows on me,
You'd have enough.

DESDEMONA: Alas! she has no speech.

IAGO: In faith, too much;
I find it still when I have list to sleep:
Marry, before your ladyship, I grant,
She puts her tongue a little in her heart,
And chides with thinking.
EMILIA: You have little cause to say so.
IAGO: Come on, come on; you are pictures out of doors,
Bells in your parlours, wild cats in your kitchens,
Saints in your injuries, devils being offended,
Players in your housewifery, and housewives in your beds.
DESDEMONA: O! fie upon thee, slanderer.
IAGO: Nay, it is true, or else I am a Turk:
You rise to play and go to bed to work.
EMILIA: You shall not write my praise.
IAGO: No, let me not.
DESDEMONA: What wouldst thou write of me, if thou shouldst praise me?
IAGO: O gentle lady, do not put me to 't,
For I am nothing if not critical.
DESDEMONA: Come on; assay. There's one gone to the harbour?
IAGO: Ay, madam.
DESDEMONA: I am not merry, but I do beguile
The thing I am by seeming otherwise.
Come, how wouldst thou praise me?
IAGO: I am about it; but indeed my invention
Comes from my pate as birdlime does from frize;[10]
It plucks out brains and all: but my muse labours,
And thus she is deliver'd.
If she be fair and wise, fairness and wit,
The one's for use, the other useth it.
DESDEMONA: Well prais'd! How if she be black and witty?
IAGO: If she be black, and thereto have a wit,
She'll find a white that shall her blackness fit.

---

[10] [*frize:* rough woolen cloth.]

DESDEMONA: Worse and worse.

EMILIA: How if fair and foolish?

IAGO: She never yet was foolish that was fair,
For even her folly help'd her to an heir.

DESDEMONA: These are old fond paradoxes to make fools laugh i' the alehouse. What miserable praise hast thou for her that's foul and foolish?

IAGO: There's none so foul and foolish thereunto
But does foul pranks which fair and wise ones do.

DESDEMONA: O heavy ignorance! thou praisest the worst best. But what praise couldst thou bestow on a deserving woman indeed, one that, in the authority of her merit, did justly put on the vouch of very malice itself?

IAGO: She that was ever fair and never proud,
Had tongue at will and yet was never loud,
Never lack'd gold and yet went never gay,
Fled from her wish and yet said "Now I may":
She that being anger'd, her revenge being nigh,
Bade her wrong stay and her displeasure fly:
She that in wisdom never was so frail
To change the cod's head for the salmon's tail:
She that could think and ne'er disclose her mind,
See suitors following and not look behind:
She was a wight, if ever such wight were,—

DESDEMONA: To do what?

IAGO: To suckle fools and chronicle small beer.

DESDEMONA: O most lame and impotent conclusion! Do not learn of him, Emilia, though he be thy husband. How say you, Cassio? is he not a most profane and liberal[11] counsellor?

CASSIO: He speaks home, madam; you may relish him more in the soldier than in the scholar.

---

[11] [*liberal:* licentious.]

IAGO [*aside*]: He takes her by the palm; ay, well said, whisper; with as little a web as this will I ensnare as great a fly as Cassio. Ay, smile upon her, do; I will gyve thee in thine own courtship. You say true, 'tis so, indeed. If such tricks as these strip you out of your lieutenantry, it had been better you had not kissed your three fingers so oft, which now again you are most apt to play the sir in. Very good; well kissed! an excellent courtesy! 'tis so, indeed. Yet again your fingers to your lips? would they were clyster-pipes for your sake! [*A trumpet heard.*] The Moor! I know his trumpet.

CASSIO: 'Tis truly so.

DESDEMONA: Let's meet him and receive him.

CASSIO: Lo! where he comes.

Enter OTHELLO *and* Attendants.

OTHELLO: O my fair warrior!
DESDEMONA: My dear Othello!
OTHELLO: It gives me wonder great as my content
To see you here before me. O my soul's joy!
If after every tempest come such calms,
May the winds blow till they have waken'd death!
And let the labouring bark climb hills of seas
Olympus-high, and duck again as low
As hell's from heaven! If it were now to die,
'Twere now to be most happy, for I fear
My soul hath her content so absolute
That not another comfort like to this
Succeeds in unknown fate.
DESDEMONA: The heavens forbid
But that our loves and comforts should increase
Even as our days do grow!
OTHELLO: Amen to that, sweet powers!
I cannot speak enough of this content;
It stops me here; it is too much of joy:

And this, and this, [*kissing her*] the greatest discords be
That e'er our hearts shall make!

IAGO [*aside*]: O! you are well tun'd now,
But I'll set down the pegs that make this music,
As honest as I am.

OTHELLO: Come, let us to the castle.
News, friends; our wars are done, the Turks are drown'd.
How does my old acquaintance of this isle?
Honey, you shall be well desir'd in Cyprus;
I have found great love amongst them. O my sweet,
I prattle out of fashion, and I dote
In mine own comforts. I prithee, good Iago,
Go to the bay and disembark my coffers.
Bring thou the master to the citadel;
He is a good one, and his worthiness
Does challenge much respect. Come, Desdemona,
Once more well met at Cyprus.

Exeunt OTHELLO *and* DESDEMONA [*with* Attendants, *etc.*].

IAGO: Do thou meet me presently at the harbour. Come hither. If thou be'st valiant, as they say base men being in love have then a nobility in their natures more than is native to them, list me. The lieutenant to-night watches on the court of guard: first, I must tell thee this, Desdemona is directly in love with him.

RODERIGO: With him! why, 'tis not possible.

IAGO: Lay thy finger thus, and let thy soul be instructed. Mark me with what violence she first loved the Moor but for bragging and telling her fantastical lies; and will she love him still for prating? let not thy discreet heart think it. Her eye must be fed; and what delight shall she have to look on the devil? When the blood is made dull with the act of sport, there should be, again to inflame it, and to give satiety a fresh appetite, loveliness in favour, sympathy in years, manners, and beauties; all which the Moor is defective in. Now, for want of these required conveniences, her delicate tenderness will find itself abused, begin to heave the gorge, disrelish and abhor the Moor; very nature will instruct her in it, and compel her to some second choice. Now, sir, this granted, as it is a most pregnant and unforced position, who stands so eminently in the degree of this fortune as Cassio does? a knave very voluble, no further conscionable than in putting on the mere form of civil and humane seeming, for the better compassing of his salt and most hidden loose affection? why, none; why, none: a slipper and subtle knave, a finder-out of occasions, that has an eye can stamp and counterfeit advantages, though true advantage never present itself; a devilish knave! Besides, the knave is handsome, young, and hath all those requisites in him that folly and green minds look after; a pestilent complete knave! and the woman hath found him already.

RODERIGO: I cannot believe that in her; she is full of most blessed condition.

IAGO: Blessed fig's end! the wine she drinks is made of grapes; if she had been blessed she would never have loved the Moor; blessed pudding! Didst thou not see her paddle with the palm of his hand? didst not mark that?

RODERIGO: Yes, that I did; but that was but courtesy.

IAGO: Lechery, by this hand! an index and obscure prologue to the history of lust and foul thoughts. They met so near with their lips, that their breaths embraced together. Villainous thoughts, Roderigo! When these mutualities so marshal the way, hard at hand comes the master and main exercise, the incorporate conclusion. Pish! But, sir, be you ruled by me: I have brought you from Venice. Watch you to-night; for the command, I'll lay't upon you: Cassio knows you not. I'll not be far from you: do you find some occasion to anger Cassio, either by speaking too loud, or tainting his discipline; or from what other course you please, which the time shall more favourably minister.

RODERIGO: Well.

IAGO: Sir, he is rash and very sudden in choler, and haply may strike at you: provoke him, that he may; for even out of that will I cause these of Cyprus to mutiny, whose qualification shall come into no true taste again but by the displanting of Cassio. So shall you have a shorter journey to your desires by the means I shall then have to prefer them; and the impediment most profitably removed, without the which there were no expectation of our prosperity.

RODERIGO: I will do this, if I can bring it to any opportunity.

IAGO: I warrant thee. Meet me by and by at the citadel: I must fetch his necessaries ashore. Farewell.

RODERIGO: Adieu. *Exit.*

IAGO: That Cassio loves her, I do well believe it;
That she loves him, 'tis apt, and of great credit:
The Moor, howbeit that I endure him not,
Is of a constant, loving, noble nature;
And I dare think he'll prove to Desdemona
A most dear husband. Now, I do love her too;
Not out of absolute lust,—though peradventure
I stand accountant for as great a sin,—
But partly led to diet my revenge,
For that I do suspect the lusty Moor
Hath leap'd into my seat; the thought whereof
Doth like a poisonous mineral gnaw my inwards;
And nothing can or shall content my soul
Till I am even'd with him, wife for wife;
Or failing so, yet that I put the Moor
At least into a jealousy so strong
That judgment cannot cure. Which thing to do,
If this poor trash of Venice, whom I trace
For his quick hunting, stand the putting-on,
I'll have our Michael Cassio on the hip;
Abuse him to the Moor in the rank garb,—
For I fear Cassio with my night-cap too,—
Make the Moor thank me, love me, and reward me

For making him egregiously an ass
And practising upon his peace and quiet
Even to madness. 'Tis here, but yet confus'd:
Knavery's plain face is never seen till us'd. *Exit.*

SCENE TWO—*A Street.*

Enter OTHELLO'*s* HERALD, *with a proclamation.* [*People following.*]

HERALD: It is Othello's pleasure, our noble and valiant general, that, upon certain tidings now arrived, importing the mere perdition of the Turkish fleet, every man put himself into triumph; some to dance, some to make bonfires, each man to what sport and revels his addiction leads him; for, besides these beneficial news, it is the celebration of his nuptial. So much was his pleasure should be proclaimed. All offices are open, and there is full liberty of feasting from this present hour of five till the bell have told eleven. Heaven bless the isle of Cyprus and our noble general Othello! *Exeunt.*

SCENE THREE—*A Hall in the Castle.*

Enter OTHELLO, DESDEMONA, CASSIO, *and* Attendants.

OTHELLO: Good Michael, look you to the guard to-night:
Let's teach ourselves that honourable stop,
Not to outsport discretion.
CASSIO: Iago hath direction what to do;
But, notwithstanding, with my personal eye
Will I look to 't.
OTHELLO: Iago is most honest.
Michael, good night; to-morrow with your earliest
Let me have speech with you.
[*To* DESDEMONA.] Come, my dear love,

The purchase made, the fruits are to ensue;
That profit's yet to come 'twixt me and you.
Good night. *Exit [with* DESDEMONA *and* Attendants].

Enter IAGO.

CASSIO: Welcome, Iago; we must to the watch.

IAGO: Not this hour, lieutenant; 'tis not yet ten o' the clock. Our general cast us thus early for the love of his Desdemona, who let us not therefore blame; he hath not yet made wanton the night with her, and she is sport for Jove.

CASSIO: She's a most exquisite lady.

IAGO: And, I'll warrant her, full of game.

CASSIO: Indeed, she is a most fresh and delicate creature.

IAGO: What an eye she has! methinks it sounds a parley of provocation.

CASSIO: An inviting eye; and yet methinks right modest.

IAGO: And when she speaks, is it not an alarum to love?

CASSIO: She is indeed perfection.

IAGO: Well, happiness to their sheets! Come, lieutenant, I have a stoup of wine, and here without are a brace of Cyprus gallants that would fain have a measure to the health of black Othello.

CASSIO: Not to-night, good Iago: I have very poor and unhappy brains for drinking: I could well wish courtesy would invent some other custom of entertainment.

IAGO: O! they are our friends; but one cup: I'll drink for you.

CASSIO: I have drunk but one cup to-night, and that was craftily qualified too, and, behold, what innovation it makes here: I am unfortunate in the infirmity, and dare not task my weakness with any more.

IAGO: What, man! 'tis a night of revels; the gallants desire it.

CASSIO: Where are they?

IAGO: Here at the door; I pray you, call them in.

CASSIO: I'll do 't; but it dislikes me. *Exit.*

IAGO: If I can fasten but one cup upon him,
With that which he hath drunk to-night already,
He'll be as full of quarrel and offence
As my young mistress' dog. Now, my sick fool Roderigo,
Whom love has turn'd almost the wrong side out,
To Desdemona hath to-night carous'd
Potations pottle-deep; and he's to watch.
Three lads of Cyprus, noble swelling spirits,
That hold their honours in a wary distance,
The very elements of this warlike isle,
Have I to-night fluster'd with flowing cups,
And they watch too. Now, 'mongst this flock of drunkards,
Am I to put our Cassio in some action
That may offend the isle. But here they come.
If consequence do but approve my dream,
My boat sails freely, both with wind and stream.

Enter CASSIO, MONTANO, *and* GENTLEMEN. [Servants *following with wine.*]

CASSIO: 'Fore God, they have given me a rouse already.
MONTANO: Good faith, a little one; not past a pint, as I am a soldier.
IAGO: Some wine, ho! [*Sings.*]

"And let me the canakin clink, clink;
And let me the canakin clink:
A soldier's a man;
A life's but a span;
Why then let a soldier drink."

Some wine, boys!
CASSIO: 'Fore God, an excellent song.
IAGO: I learned it in England, where indeed they are most potent in potting; your Dane, your German, and your swag-bellied Hollander,—drink, ho!—are nothing to your English.

CASSIO: Is your Englishman so expert in his drinking?

IAGO: Why, he drinks you with facility your Dane dead drunk; he sweats not to overthrow your Almain; he gives your Hollander a vomit ere the next pottle can be filled.

CASSIO: To the health of our general!

MONTANO: I am for it, lieutenant; and I'll do you justice.

IAGO: O sweet England! [*Sings.*]

"King Stephen was a worthy peer,
  His breeches cost him but a crown;
He held them sixpence all too dear,
  With that he call'd the tailor lown.
He was a wight of high renown,
  And thou art but of low degree:
'Tis pride that pulls the country down,
  Then take thine auld cloak about thee."

Some wine, ho!

CASSIO: Why, this is a more exquisite song than the other.

IAGO: Will you hear 't again?

CASSIO: No; for I hold him to be unworthy of his place that does those things. Well, God's above all; and there be souls must be saved, and there be souls must not be saved.

IAGO: It's true, good lieutenant.

CASSIO: For mine own part,—no offence to the general, nor any man of quality,—I hope to be saved.

IAGO: And so do I too, lieutenant.

CASSIO: Ay; but, by your leave, not before me; the lieutenant is to be saved before the ancient. Let's have no more of this; let's to our affairs. God forgive us our sins! Gentlemen, let's look to our business. Do not think, gentlemen, I am drunk: this is my ancient; this is my right hand, and this is my left hand. I am not drunk now; I can stand well enough, and speak well enough.

ALL: Excellent well.

CASSIO: Why, very well, then; you must not think then that I am drunk. *Exit.*

MONTANO: To the platform, masters; come, let's set the watch.

IAGO: You see this fellow that is gone before;
He is a soldier fit to stand by Caesar
And give direction; and do but see his vice;
'Tis to his virtue a just equinox,
The one as long as the other; 'tis pity of him.
I fear the trust Othello puts him in,
On some odd time of his infirmity,
Will shake this island.

MONTANO: But is he often thus?

IAGO: 'Tis evermore the prologue to his sleep:
He'll watch the horologe a double set,[12]
If drink rock not his cradle.

MONTANO: It were well
The general were put in mind of it.
Perhaps he sees it not; or his good nature
Prizes the virtue that appears in Cassio,
And looks not on his evils. Is not this true?

Enter RODERIGO.

IAGO [*aside to him*]: How now, Roderigo!
I pray you, after the lieutenant; go. *Exit* RODERIGO.

MONTANO: And 'tis great pity that the noble Moor
Should hazard such a place as his own second
With one of an ingraft infirmity;
It were an honest action to say
So to the Moor.

IAGO: Not I, for this fair island:
I do love Cassio well, and would do much
To cure him of this evil. But hark! what noise?

---

[12] [*watch the horologe a double set:* stay up to see the horologe (clock) go around twice.]

[*Cry*] *within.* "Help! Help!"

Enter CASSIO, *pursuing* RODERIGO.

CASSIO: You rogue, you rascal!
MONTANO: What's the matter, lieutenant?
CASSIO: A knave teach me my duty!
I'll beat the knave into a twiggen bottle.
RODERIGO: Beat me!
CASSIO: Dost thou prate, rogue?
[*Striking* RODERIGO.]
MONTANO [*staying him*]: Nay, good lieutenant;
I pray you, sir, hold your hand.
CASSIO: Let me go, sir,
Or I'll knock you o'er the mazzard.
MONTANO: Come, come; you're drunk.
CASSIO: Drunk! *They fight.*
IAGO [*aside to* RODERIGO]: Away, I say! go out, and cry a mutiny. *Exit* RODERIGO.
Nay, good lieutenant! God's will, gentlemen!
Help, ho! Lieutenant! sir! Montano! sir!
Help, masters! Here's a goodly watch indeed! *Bell rings.*
Who's that that rings the bell? *Diablo,* ho!
The town will rise: God's will! lieutenant, hold!
You will be sham'd for ever.

Enter OTHELLO *and* Attendants.

OTHELLO: What is the matter here?
MONTANO: 'Zounds! I bleed still; I am hurt to the death.
OTHELLO: Hold, for your lives!
IAGO: Hold, ho, lieutenant! Sir! Montano! gentlemen!
Have you forgot all sense of place and duty?
Hold! the general speaks to you; hold for shame!
OTHELLO: Why, how now, ho! from whence ariseth this?
Are we turn'd Turks, and to ourselves do that
Which heaven hath forbid the Ottomites?

For Christian shame put by this barbarous brawl;
He that stirs next to carve for his own rage
Holds his soul light; he dies upon his motion.
Silence that dreadful bell! it frights the isle
From her propriety. What is the matter, masters?
Honest Iago, that look'st dead with grieving,
Speak, who began this? on thy love, I charge thee.

IAGO: I do not know; friends all but now, even now,
In quarter and in terms like bride and groom
Devesting them for bed; and then, but now,—
As if some planet had unwitted men,—
Swords out, and tilting one at other's breast,
In opposition bloody. I cannot speak
Any beginning to this peevish odds,
And would in action glorious I had lost
Those legs that brought me to a part of it!

OTHELLO: How comes it, Michael, you are thus forgot?

CASSIO: I pray you, pardon me; I cannot speak.

OTHELLO: Worthy Montano, you were wont be civil;
The gravity and stillness of your youth
The world hath noted, and your name is great
In mouths of wisest censure: what's the matter,
That you unlace your reputation thus
And spend your rich opinion for the name
Of a night-brawler? give me answer to it.

MONTANO: Worthy Othello, I am hurt to danger;
Your officer, Iago, can inform you,
While I spare speech, which something now offends me,
Of all that I do know; nor know I aught
By me that's said or done amiss this night,
Unless self-charity be sometimes a vice,
And to defend ourselves it be a sin
When violence assails us.

OTHELLO: Now, by heaven,
My blood begins my safer guides to rule,
And passion, having my best judgment collied,

Assays to lead the way. If I once stir,
Or do but lift this arm, the best of you
Shall sink in my rebuke. Give me to know
How this foul rout began, who set it on;
And he that is approv'd in this offence,
Though he had twinn'd with me—both at a birth—
Shall lose me. What! in a town of war,
Yet wild, the people's hearts brimful of fear,
To manage private and domestic quarrel,
In night, and on the court and guard of safety!
'Tis monstrous. Iago, who began 't?

MONTANO: If partially affin'd, or leagu'd in office,
Thou dost deliver more or less than truth,
Thou art no soldier.

IAGO: Touch me not so near;
I had rather have this tongue cut from my mouth
Than it should do offence to Michael Cassio;
Yet, I persuade myself, to speak the truth
Shall nothing wrong him. Thus it is, general.
Montano and myself being in speech,
There comes a fellow crying out for help,
And Cassio following with determin'd sword
To execute upon him. Sir, this gentleman
Steps in to Cassio, and entreats his pause;
Myself the crying fellow did pursue,
Lest by his clamour, as it so fell out,
The town might fall in fright; he, swift of foot,
Outran my purpose, and I return'd the rather
For that I heard the clink and fall of swords,
And Cassio high in oath, which till to-night
I ne'er might say before. When I came back,—
For this was brief,—I found them close together,
At blow and thrust, even as again they were
When you yourself did part them.
More of this matter can I not report:
But men are men; the best sometimes forget:

Though Cassio did some little wrong to him,
As men in rage strike those that wish them best,
Yet, surely Cassio, I believe, receiv'd
From him that fled some strange indignity,
Which patience could not pass.

OTHELLO: I know, Iago,
Thy honesty and love doth mince this matter,
Making it light to Cassio. Cassio, I love thee;
But never more be officer of mine.

Enter DESDEMONA, *attended.*

Look, if my gentle love be not rais'd up!
[*To* CASSIO.] I'll make thee an example.

DESDEMONA: What's the matter?

OTHELLO: All's well now, sweeting; come away to bed.
Sir, for your hurts, myself will be your surgeon.
Lead him off. [MONTANO *is led off.*]
Iago, look with care about the town,
And silence those whom this vile brawl distracted.
Come, Desdemona; 'tis the soldiers' life,
To have their balmy slumbers wak'd with strife.
Exit [*with* DESDEMONA *and* Attendants].

IAGO: What! are you hurt, lieutenant?

CASSIO: Ay; past all surgery.

IAGO: Marry, heaven forbid!

CASSIO: Reputation, reputation, reputation! O! I have lost my reputation. I have lost the immortal part of myself, and what remains is bestial. My reputation, Iago, my reputation!

IAGO: As I am an honest man, I thought you had received some bodily wound; there is more sense in that than in reputation. Reputation is an idle and most false imposition; oft got without merit, and lost without deserving: you have lost no reputation at all, unless you repute yourself such a loser. What, man! there are ways to recover the general again; you are but now cast in his mood, a punishment more in policy than in malice; even so as one would beat his offenceless dog to affright an imperious lion. Sue to him again, and he is yours.

CASSIO: I will rather sue to be despised than to deceive so good a commander with so slight, so drunken, and so indiscreet an officer. Drunk! and speak parrot! and squabble, swagger, swear, and discourse fustian with one's own shadow! O thou invisible spirit of wine! if thou hast no name to be known by, let us call thee devil!

IAGO: What was he that you followed with your sword? What had he done to you?

CASSIO: I know not.

IAGO: Is 't possible?

CASSIO: I remember a mass of things, but nothing distinctly; a quarrel, but nothing wherefore. O God! that men should put an enemy in their mouths to steal away their brains; that we should, with joy, pleasance, revel, and applause, transform ourselves into beasts.

IAGO: Why, but you are now well enough; how came you thus recovered?

CASSIO: It hath pleased the devil drunkenness to give place to the devil wrath; one unperfectness shows me another, to make me frankly despise myself.

IAGO: Come, you are too severe a moraler. As the time, the place, and the condition of this country stands, I could heartily wish this had not befallen, but since it is as it is, mend it for your own good.

CASSIO: I will ask him for my place again; he shall tell me I am a drunkard! Had I as many mouths as Hydra, such an answer would stop them all. To be now a sensible man, by and by a fool, and presently a beast! O strange! Every inordinate cup is unblessed and the ingredient is a devil.

IAGO: Come, come; good wine is a good familiar creature if it be well used; exclaim no more against it. And, good lieutenant, I think you think I love you.

CASSIO: I have well approved it, sir. I drunk!

IAGO: You or any man living may be drunk at some time, man. I'll tell you what you shall do. Our general's wife is now the general: I may say so in this respect, for that he hath devoted and given up himself to the contemplation, mark, and denotement of her parts and graces: confess yourself freely to her; importune her; she'll help to put you in your place again. She is of so free, so kind, so apt, so blessed a disposition, that she holds it a vice in her goodness not to do more than she is requested. This broken joint between you and her husband entreat her to splinter; and my fortunes against any lay worth naming, this crack of your love shall grow stronger than it was before.

CASSIO: You advise me well.

IAGO: I protest, in the sincerity of love and honest kindness.

CASSIO: I think it freely; and betimes in the morning I will beseech the virtuous Desdemona to undertake for me. I am desperate of my fortunes if they check me here.

IAGO: You are in the right. Good night, lieutenant; I must to the watch.

CASSIO: Good night, honest Iago! *Exit* CASSIO.

IAGO: And what's he then that says I play the villain?
When this advice is free I give and honest,
Probal to thinking and indeed the course
To win the Moor again? For 'tis most easy
The inclining Desdemona to subdue
In any honest suit; she's fram'd as fruitful
As the free elements. And then for her
To win the Moor, were 't to renounce his baptism,
All seals and symbols of redeemed sin,
His soul is so enfetter'd to her love,
That she may make, unmake, do what she list,
Even as her appetite shall play the god
With his weak function. How am I then a villain
To counsel Cassio to this parallel course,
Directly to his good? Divinity of hell!

When devils will the blackest sins put on,
They do suggest at first with heavenly shows,
As I do now; for while this honest fool
Plies Desdemona to repair his fortunes,
And she for him pleads strongly to the Moor,
I'll pour this pestilence into his ear
That she repeals him for her body's lust;
And, by how much she strives to do him good,
She shall undo her credit with the Moor.
So will I turn her virtue into pitch,
And out of her own goodness make the net
That shall enmesh them all.

Enter RODERIGO.

How now, Roderigo!

RODERIGO: I do follow here in the chase, not like a hound that hunts, but one that fills up the cry. My money is almost spent; I have been to-night exceedingly well cudgelled; and I think the issue will be, I shall have so much experience for my pains; and so, with no money at all and a little more wit, return again to Venice.

IAGO: How poor are they that have not patience!
What wound did ever heal but by degrees?
Thou know'st we work by wit and not by witchcraft,
And wit depends on dilatory time.
Does 't not go well? Cassio hath beaten thee,
And thou by that small hurt hast cashiered Cassio.
Though other things grow fair against the sun,
Yet fruits that blossom first will first be ripe:
Content thyself awhile. By the mass, 'tis morning;
Pleasure and action make the hours seem short.
Retire thee; go where thou art billeted:
Away, I say; thou shalt know more hereafter:
Nay, get thee gone. *Exit* RODERIGO.
Two things are to be done,
My wife must move for Cassio to her mistress;

I'll set her on;
Myself the while to draw the Moor apart,
And bring him jump when he may Cassio find
Soliciting his wife: ay, that's the way:
Dull not device by coldness and delay. *Exit.*

ACT III

SCENE ONE—*Before the Castle.*

Enter CASSIO *with* MUSICIANS.

CASSIO: Masters, play here, I will content your pains;
Something that's brief; and bid "Good morrow, general."

They play, and enter the CLOWN.

CLOWN: Why, masters, have your instruments been in Naples, that they speak i' the nose thus?

FIRST MUSICIAN: How, sir, how?

CLOWN: Are these, I pray you, wind-instruments?

FIRST MUSICIAN: Ay, marry, are they, sir.

CLOWN: O! thereby hangs a tail.

FIRST MUSICIAN: Whereby hangs a tale, sir?

CLOWN: Marry, sir, by many a wind-instrument that I know. But, masters, here's money for you; and the general so likes your music, that he desires you, for love's sake, to make no more noise with it.

FIRST MUSICIAN: Well, sir, we will not.

CLOWN: If you have any music that may not be heard, to 't again; but, as they say, to hear music the general does not greatly care.

FIRST MUSICIAN: We have none such, sir.

CLOWN: Then put up your pipes in your bag, for I'll away. Go; vanish into air; away! *Exeunt* MUSICIANS.

CASSIO: Dost thou hear, mine honest friend?

CLOWN: No, I hear not your honest friend; I hear you.

CASSIO: Prithee, keep up thy quillets. There's a poor piece of gold for thee. If the gentlewoman that attends the general's wife be stirring, tell her there's one Cassio entreats her a little favour of speech: wilt thou do this?

CLOWN: She is stirring, sir: if she will stir hither, I shall seem to notify unto her.

CASSIO: Do, good my friend. *Exit* CLOWN.

Enter IAGO.

In happy time, Iago.

IAGO: You have not been a-bed, then?

CASSIO: Why, no; the day had broke
Before we parted. I have made bold, Iago,
To send in to your wife; my suit to her
Is, that she will to virtuous Desdemona
Procure me some access.

IAGO: I'll send her to you presently;
And I'll devise a mean to draw the Moor
Out of the way, that your converse and business
May be more free.

CASSIO: I humbly thank you for 't. [*Exit* IAGO.]
I never knew
A Florentine more kind and honest.

Enter EMILIA.

EMILIA: Good morrow, good lieutenant: I am sorry
For your displeasure; but all will soon be well.
The general and his wife are talking of it,
And she speaks for you stoutly: the Moor replies
That he you hurt is of great fame in Cyprus
And great affinity, and that in wholesome wisdom
He might not but refuse you; but he protests he loves you,

And needs no other suitor but his likings
To take the saf'st occasion by the front
To bring you in again.

CASSIO: Yet, I beseech you,
If you think fit, or that it may be done,
Give me advantage of some brief discourse
With Desdemona alone.

EMILIA: Pray you, come in:
I will bestow you where you shall have time
To speak your bosom freely.

CASSIO: I am much bound to you.
[*Exeunt.*]

SCENE TWO—*A Room in the Castle.*

Enter OTHELLO, IAGO, *and* GENTLEMEN.

OTHELLO: These letters give, Iago, to the pilot,
And by him do my duties to the senate;
That done, I will be walking on the works;
Repair there to me.

IAGO: Well, my good lord, I'll do 't.

OTHELLO: This fortification, gentlemen, shall we see 't?

GENTLEMEN: We'll wait upon your lordship. *Exeunt.*

SCENE THREE—*The Garden of the Castle.*

Enter DESDEMONA, CASSIO, *and* EMILIA.

DESDEMONA: Be thou assur'd, good Cassio, I will do
All my abilities in thy behalf.

EMILIA: Good madam, do: I warrant it grieves my husband,
As if the case were his.

DESDEMONA: O! that's an honest fellow. Do not doubt, Cassio,
But I will have my lord and you again
As friendly as you were.

CASSIO: Bounteous madam,
Whatever shall become of Michael Cassio,
He's never anything but your true servant.
DESDEMONA: I know 't; I thank you. You do love my lord;
You have known him long; and be you well assur'd
He shall in strangeness stand no further off
Than in a politic distance.
CASSIO: Ay, but, lady,
That policy may either last so long,
Or feed upon such nice and waterish diet,
Or breed itself so out of circumstance,
That, I being absent and my place supplied,
My general will forget my love and service.
DESDEMONA: Do not doubt that; before Emilia here
I give thee warrant of thy place. Assure thee,
If I do vow a friendship, I'll perform it
To the last article; my lord shall never rest;
I'll watch him tame, and talk him out of patience;
His bed shall seem a school, his board a shrift;[13]
I'll intermingle everything he does
With Cassio's suit. Therefore be merry, Cassio;
For thy solicitor shall rather die
Than give thy cause away.

Enter OTHELLO *and* IAGO [*at a distance*].

EMILIA: Madam, here comes my lord.
CASSIO: Madam, I'll take my leave.
DESDEMONA: Why, stay, and hear me speak.
CASSIO: Madam, not now; I am very ill at ease,
Unfit for mine own purposes.
DESDEMONA: Well, do your discretion. *Exit* CASSIO.
IAGO: Ha! I like not that.
OTHELLO: What dost thou say?

---

[13] [*shrift:* confessional.]

IAGO: Nothing, my lord: or if—I know not what.
OTHELLO: Was not that Cassio parted from my wife?
IAGO: Cassio, my lord? No, sure, I cannot think it
That he would steal away so guilty-like,
Seeing you coming.
OTHELLO: I do believe 'twas he.
DESDEMONA: How now, my lord!
I have been talking with a suitor here,
A man that languishes in your displeasure.
OTHELLO: Who is 't you mean?
DESDEMONA: Why, your lieutenant, Cassio. Good my lord,
If I have any grace or power to move you,
His present reconciliation take;
For if he be not one that truly loves you,
That errs in ignorance and not in cunning,
I have no judgment in an honest face.
I prithee call him back.
OTHELLO: Went he hence now?
DESDEMONA: Ay, sooth; so humbled,
That he haft left part of his grief with me,
To suffer with him. Good love, call him back.
OTHELLO: Not now, sweet Desdemona; some other time.
DESDEMONA: But shall 't be shortly?
OTHELLO: The sooner, sweet, for you.
DESDEMONA: Shall 't be to-night at supper?
OTHELLO: No, not to-night.
DESDEMONA: To-morrow dinner then?
OTHELLO: I shall not dine at home;
I meet the captains at the citadel.
DESDEMONA: Why then, to-morrow night; or Tuesday morn;
On Tuesday noon, or night; or Wednesday morn:
I prithee name the time, but let it not
Exceed three days: in faith, he's penitent;
And yet his trespass, in our common reason,—
Save that they say, the wars must make examples
Out of their best,—is not almost a fault

To incur a private check. When shall he come?
Tell me, Othello: I wonder in my soul,
What you could ask me that I should deny,
Or stand so mammering on. What! Michael Cassio,
That came a wooing with you, and so many a time,
When I have spoke of you dispraisingly,
Hath ta'en your part; to have so much to do
To bring him in! Trust me, I could do much—

OTHELLO: Prithee, no more; let him come when he will;
I will deny thee nothing.

DESDEMONA: Why, this is not a boon;
'Tis as I should entreat you wear your gloves,
Or feed on nourishing dishes, or keep you warm,
Or sue to you to do a peculiar profit
To your own person; nay, when I have a suit
Wherein I mean to touch your love indeed,
It shall be full of poise and difficult weight,
And fearful to be granted.

OTHELLO: I will deny thee nothing:
Whereon, I do beseech thee, grant me this,
To leave me but a little to myself.

DESDEMONA: Shall I deny you? no: farewell, my lord.

OTHELLO: Farewell, my Desdemona: I'll come to thee straight.

DESDEMONA: Emilia, come. Be as your fancies teach you;
Whate'er you be, I am obedient. *Exit* [*with* EMILIA].

OTHELLO: Excellent wretch! Perdition catch my soul
But I do love thee! and when I love thee not,
Chaos is come again.

IAGO: My noble lord,—

OTHELLO: What dost thou say, Iago?

IAGO: Did Michael Cassio, when you woo'd my lady,
Know of your love?

OTHELLO: He did, from first to last: why dost thou ask?

IAGO: But for a satisfaction of my thought;
No further harm.

OTHELLO: Why of thy thought, Iago?

IAGO: I did not think he had been acquainted with her.
OTHELLO: O! yes; and went between us very oft.
IAGO: Indeed!
OTHELLO: Indeed! ay, indeed; discern'st thou aught in that?
Is he not honest?
IAGO: Honest, my lord?
OTHELLO: Honest! ay, honest.
IAGO: My lord, for aught I know.
OTHELLO: What dost thou think?
IAGO: Think, my lord!
OTHELLO: Think, my lord!
By heaven, he echoes me,
As if there were some monster in his thought
Too hideous to be shown. Thou dost mean something:
I heard thee say but now, thou lik'dst not that,
When Cassio left my wife; what didst not like?
And when I told thee he was of my counsel
In my whole course of wooing, thou criedst, "Indeed!"
And didst contract and purse thy brow together,
As if thou then hadst shut up in thy brain
Some horrible conceit.[14] If thou dost love me,
Show me thy thought.
IAGO: My lord, you know I love you.
OTHELLO: I think thou dost;
And, for I know thou art full of love and honesty,
And weigh'st thy words before thou giv'st them breath,
Therefore these stops of thine fright me the more;
For such things in a false disloyal knave
Are tricks of custom, but in a man that's just
They are close delations,[15] working from the heart
That passion cannot rule.
IAGO: For Michael Cassio,
I dare be sworn I think that he is honest.

---

[14] [*conceit:* idea.]

[15] [*close delations:* covert, involuntary accusations.]

OTHELLO: I think so too.
IAGO: Men should be what they seem;
Or those that be not, would they might seem none!
OTHELLO: Certain, men should be what they seem.
IAGO: Why then, I think Cassio's an honest man.
OTHELLO: Nay, yet there's more in this.
I pray thee, speak to me as to thy thinkings,
As thou dost ruminate, and give thy worst of thoughts
The worst of words.
IAGO: Good my lord, pardon me;
Though I am bound to every act of duty,
I am not bound to that all slaves are free to.
Utter my thoughts? Why, say they are vile and false;
As where's that palace whereinto foul things
Sometimes intrude not? who has a breast so pure
But some uncleanly apprehensions
Keep leets and law-days, and in session sit
With meditations lawful?
OTHELLO: Thou dost conspire against thy friend, Iago,
If thou but think'st him wrong'd, and mak'st his ear
A stranger to thy thoughts.
IAGO: I do beseech you,
Though I perchance am vicious in my guess,—
As, I confess, it is my nature's plague
To spy into abuses, and oft my jealousy
Shapes faults that are not,—that your wisdom yet,
From one that so imperfectly conceits,
Would take no notice, nor build yourself a trouble
Out of his scattering and unsure observance.
It were not for your quiet nor your good,
Nor for my manhood, honesty, or wisdom,
To let you know my thoughts.
OTHELLO: What dost thou mean?
IAGO: Good name in man and woman, dear my lord,
Is the immediate jewel of their souls:
Who steals my purse steals trash; 'tis something, nothing;

'Twas mine, 'tis his, and has been slave to thousands;
But he that filches from me my good name
Robs me of that which not enriches him,
And makes me poor indeed.

OTHELLO: By heaven, I'll know thy thoughts.

IAGO: You cannot, if my heart were in your hand;
Nor shall not, whilst 'tis in my custody.

OTHELLO: Ha!

IAGO: O! beware, my lord, of jealousy;
It is the green-ey'd monster which doth mock
The meat it feeds on; that cuckold lives in bliss
Who, certain of his fate, loves not his wronger;
But, O! what damned minutes tells he o'er
Who dotes, yet doubts; suspects, yet strongly loves!

OTHELLO: O misery!

IAGO: Poor and content is rich, and rich enough,
But riches fineless is as poor as winter
To him that ever fears he shall be poor.
Good heaven, the souls of all my tribe defend
From jealousy.

OTHELLO: Why, why is this?
Think'st thou I'd make a life of jealousy,
To follow still the changes of the moon
With fresh suspicions? No; to be once in doubt
Is once to be resolv'd. Exchange me for a goat
When I shall turn the business of my soul
To such exsufflicate[16] and blown surmises,
Matching thy inference. 'Tis not to make me jealous
To say my wife is fair, feeds well, loves company,
Is free of speech, sings, plays, and dances well;
Where virtue is, these are more virtuous:
Nor from mine own weak merits will I draw
The smallest fear, or doubt of her revolt;
For she had eyes, and chose me. No, Iago;

---

[16] [*exsufflicate:* puffed up, empty, or possibly, contemptible.]

I'll see before I doubt; when I doubt, prove;
And, on the proof, there is no more but this,
Away at once with love or jealousy!

IAGO: I am glad of it; for now I shall have reason
To show the love and duty that I bear you
With franker spirit; therefore, as I am bound,
Receive it from me; I speak not yet of proof.
Look to your wife; observe her well with Cassio;
Wear your eye thus, not jealous nor secure:
I would not have your free and noble nature
Out of self-bounty be abus'd; look to 't:
I know our country disposition well;
In Venice they do let heaven see the pranks
They dare not show their husbands; their best conscience
Is not to leave 't undone, but keep 't unknown.

OTHELLO: Dost thou say so?

IAGO: She did deceive her father, marrying you:
And when she seem'd to shake and fear your looks,
She lov'd them most.

OTHELLO: And so she did.

IAGO: Why, go to, then;
She that so young could give out such a seeming,
To seel her father's eyes up close as oak,
He thought 'twas witchcraft—but I am much to blame:
I humbly do beseech you of your pardon
For too much loving you.

OTHELLO: I am bound to thee for ever.

IAGO: I see, this hath a little dash'd your spirits.

OTHELLO: Not a jot, not a jot.

IAGO: I' faith, I fear it has.
I hope you will consider what is spoke
Comes from my love. But, I do see you're mov'd;
I am to pray you not to strain my speech
To grosser issues nor to larger reach
Than to suspicion.

OTHELLO: I will not.
IAGO: Should you do so, my lord,
My speech should fall into such vile success
As my thoughts aim not at. Cassio's my worthy friend—
My lord, I see you're mov'd.
OTHELLO: No, not much mov'd:
I do not think but Desdemona's honest.
IAGO: Long live she so! and long live you to think so!
OTHELLO: And, yet, how nature erring from itself,—
IAGO: Ay, there's the point: as, to be bold with you,
Not to affect many proposed matches
Of her own clime, complexion, and degree,
Whereto, we see, in all things nature tends;
Foh! one may smell in such, a will most rank,
Foul disproportion, thoughts unnatural.
But pardon me; I do not in position
Distinctly speak of her, though I may fear
Her will, recoiling to her better judgment,
May fall to match you with her country forms
And happily repent.
OTHELLO: Farewell, farewell:
If more thou dost perceive, let me know more;
Set on thy wife to observe. Leave me, Iago.
IAGO: My lord, I take my leave. [*Going.*]
OTHELLO: Why did I marry? This honest creature, doubtless,
Sees and knows more, much more, than he unfolds.
IAGO [*returning*]: My lord, I would I might entreat your honour
To scan this thing no further; leave it to time.
Although 'tis fit that Cassio have his place,
For sure he fills it up with great ability,
Yet, if you please to hold him off awhile,
You shall by that perceive him and his means:
Note if your lady strain his entertainment
With any strong or vehement importunity;
Much will be seen in that. In the mean time,
Let me be thought too busy in my fears,

As worthy cause I have to fear I am,
And hold her free, I do beseech your honour.
OTHELLO: Fear not my government.
IAGO: I once more take my leave. *Exit.*
OTHELLO: This fellow's of exceeding honesty,
And knows all qualities, with a learned spirit,
Of human dealings; if I do prove her haggard,
Though that her jesses were my dear heartstrings,
I'd whistle her off and let her down the wind,
To prey at fortune. Haply, for I am black,
And have not those soft parts of conversation
That chamberers have, or, for I am declin'd
Into the vale of years—yet that's not much—
She's gone, I am abus'd; and my relief
Must be to loathe her. O curse of marriage!
That we can call these delicate creatures ours,
And not their appetites. I had rather be a toad,
And live upon the vapour of a dungeon,
Than keep a corner in the thing I love
For others' uses. Yet, 'tis the plague of great ones;
Prerogativ'd are they less than the base;
'Tis destiny unshunnable, like death:
Even then this forked plague is fated to us
When we do quicken.
Look! where she comes.
If she be false, O! then heaven mocks itself.
I'll not believe it!

Enter DESDEMONA *and* EMILIA.

DESDEMONA: How now, my dear Othello!
Your dinner and generous islanders
By you invited, do attend your presence.
OTHELLO: I am to blame.
DESDEMONA: Why do you speak so faintly?
Are you not well?

OTHELLO: I have a pain upon my forehead here.
DESDEMONA: Faith, that's with watching; 'twill away again:
Let me but bind it hard, within this hour
It will be well.
OTHELLO: Your napkin is too little:
[*He puts the handkerchief from him; and it drops.*]
Let it alone. Come, I'll go in with you.
DESDEMONA: I am very sorry that you are not well.
Exeunt OTHELLO *and* DESDEMONA.

EMILIA: I am glad I have found this napkin;
This was her first remembrance from the Moor;
My wayward husband hath a hundred times
Woo'd me to steal it, but she so loves the token,
For he conjur'd her she should ever keep it,
That she reserves it evermore about her
To kiss and talk to. I'll have the work ta'en out,[17]
And give 't Iago:
What he will do with it heaven knows, not I;
I nothing but to please his fantasy.

Enter IAGO.

IAGO: How now! what do you here alone?
EMILIA: Do not you chide; I have a thing for you.
IAGO: A thing for me? It is a common thing—
EMILIA: Ha!
IAGO: To have a foolish wife.
EMILIA: O! is that all? What will you give me now
For that same handkerchief?
IAGO: What handkerchief?
EMILIA: What handkerchief!
Why, that the Moor first gave to Desdemona:
That which so often you did bid me steal.

---

[17] [*work ta'en out:* embroidery copied.]

IAGO: Hast stol'n it from her?
EMILIA: No, faith; she let it drop by negligence,
And, to the advantage, I, being there, took 't up.
Look, here it is.
IAGO: A good wench; give it me.
EMILIA: What will you do with 't, that you have been so earnest
To have me filch it?
IAGO: Why, [*snatching it*] what's that to you?
EMILIA: If it be not for some purpose of import
Give 't me again; poor lady! she'll run mad
When she shall lack it.
IAGO: Be not acknown on 't; I have use for it.
Go, leave me. *Exit* EMILIA.

I will in Cassio's lodging lose this napkin,
And let him find it; trifles light as air
Are to the jealous confirmations strong
As proofs of holy writ; this may do something.
The Moor already changes with my poison:
Dangerous conceits are in their natures poisons,
Which at the first are scarce found to distaste,
But with a little act upon the blood,
Burn like the mines of sulphur. I did say so:
Look! where he comes!

Enter OTHELLO.

Not poppy, nor mandragora,
Nor all the drowsy syrups of the world,
Shall ever medicine thee to that sweet sleep
Which thou ow'dst yesterday.
OTHELLO: Ha! ha! false to me?
IAGO: Why, how now, general! no more of that.
OTHELLO: Avaunt! be gone! thou hast set me on the rack;
I swear 'tis better to be much abus'd
Than but to know 't a little.
IAGO: How now, my lord!

OTHELLO: What sense had I of her stol'n hours of lust?
I saw 't not, thought it not, it harm'd not me;
I slept the next night well, was free and merry;
I found not Cassio's kisses on her lips;
He that is robb'd, not wanting what is stol'n,
Let him not know 't and he's not robb'd at all.
IAGO: I am sorry to hear this.
OTHELLO: I had been happy, if the general camp,
Pioners and all, had tasted her sweet body,
So I had nothing known. O! now, for ever
Farewell the tranquil mind; farewell content!
Farewell the plumed troop and the big wars
That make ambition virtue! O, farewell!
Farewell the neighing steed, and the shrill trump,
The spirit-stirring drum, the ear-piercing fife,
The royal banner, and all quality,
Pride, pomp, and circumstance of glorious war!
And, O you mortal engines, whose rude throats
The immortal Jove's dread clamours counterfeit,
Farewell! Othello's occupation's gone!
IAGO: Is it possible, my lord?
OTHELLO: Villain, be sure thou prove my love a whore,
Be sure of it; give me the ocular proof;
Or, by the worth of mine eternal soul,
Thou hadst been better have been born a dog
Than answer my wak'd wrath.
IAGO: Is 't come to this?
OTHELLO: Make me to see 't; or, at the least, so prove it,
That the probation bear no hinge nor loop
To hang a doubt on; or woe upon thy life!
IAGO: My noble lord,—
OTHELLO: If thou dost slander her and torture me,
Never pray more; abandon all remorse;
On horror's head horrors accumulate;
Do deeds to make heaven weep, all earth amaz'd;
For nothing canst thou to damnation add
Greater than that.

IAGO: O grace! O heaven forgive me!
Are you a man! have you a soul or sense?
God be wi' you; take mine office. O wretched fool!
That liv'st to make thine honesty a vice.
O monstrous world! Take note, take note, O world!
To be direct and honest is not safe.
I thank you for this profit, and, from hence
I'll love no friend, sith love breeds such offence.
OTHELLO: Nay, stay; thou shouldst be honest.
IAGO: I should be wise; for honesty's a fool,
And loses that it works for.
OTHELLO: By the world,
I think my wife be honest and think she is not;
I think that thou art just and think thou art not.
I'll have some proof. Her name, that was as fresh
As Dian's visage, is now begrim'd and black
As mine own face. If there be cords or knives,
Poison or fire or suffocating streams,
I'll not endure it. Would I were satisfied!
IAGO: I see, sir, you are eaten up with passion.
I do repent me that I put it to you.
You would be satisfied?
OTHELLO: Would! nay, I will.
IAGO: And may; but how? how satisfied, my lord?
Would you, the supervisor, grossly gape on;
Behold her tupp'd?
OTHELLO: Death and damnation! O!
IAGO: It were a tedious difficulty, I think,
To bring them to that prospect; damn them then,
If ever mortal eyes do see them bolster
More than their own! What then? how then?
What shall I say? Where's satisfaction?
It is impossible you should see this,
Were they as prime as goats, as hot as monkeys,
As salt as wolves in pride, and fools as gross
As ignorance made drunk; but yet, I say,

If imputation, and strong circumstances,
Which lead directly to the door of truth,
Will give you satisfaction, you may have it.
OTHELLO: Give me a living reason she's disloyal.
IAGO: I do not like the office;
But, sith I am enter'd in this cause so far,
Prick'd to 't by foolish honesty and love,
I will go on. I lay with Cassio lately;
And, being troubled with a raging tooth,
I could not sleep.
There are a kind of men so loose of soul
That in their sleeps will mutter their affairs;
One of this kind is Cassio.
In sleep I heard him say, "Sweet Desdemona,
Let us be wary, let us hide our loves!"
And then, sir, would he gripe and wring my hand,
Cry, "O, sweet creature!" and then kiss me hard,
As if he pluck'd up kisses by the roots,
That grew upon my lips; then laid his leg
Over my thigh, and sigh'd, and kiss'd; and then
Cried, "Cursed fate, that gave thee to the Moor!"
OTHELLO: O monstrous! monstrous!
IAGO: Nay, this was but his dream.
OTHELLO: But this denoted a foregone conclusion:
'Tis a shrewd doubt, though it be but a dream.
IAGO: And this may help to thicken other proofs
That do demonstrate thinly.
OTHELLO: I'll tear her all to pieces.
IAGO: Nay, but be wise; yet we see nothing done;
She may be honest yet. Tell me but this:
Have you not sometimes seen a handkerchief
Spotted with strawberries in your wife's hand?
OTHELLO: I gave her such a one; 'twas my first gift.
IAGO: I know not that; but such a handkerchief—
I am sure it was your wife's—did I to-day
See Cassio wipe his beard with.

OTHELLO: If it be that,—
IAGO: If it be that, or any that was hers,
It speaks against her with the other proofs.
OTHELLO: O! that the slave had forty thousand lives;
One is too poor, too weak for my revenge.
Now do I see 'tis true. Look here, Iago;
All my fond love thus do I blow to heaven:
'Tis gone.
Arise, black vengeance, from the hollow hell!
Yield up, O love! thy crown and hearted throne
To tyrannous hate. Swell, bosom, with thy fraught,
For 'tis of aspics' tongues!
IAGO: Yet be content.
OTHELLO: O! blood, blood, blood!
IAGO: Patience, I say; your mind, perhaps, may change.
OTHELLO: Never, Iago. Like to the Pontic sea,
Whose icy current and compulsive course
Ne'er feels retiring ebb, but keeps due on
To the Propontic and the Hellespont,
Even so my bloody thoughts, with violent pace,
Shall ne'er look back, ne'er ebb to humble love,
Till that a capable and wide revenge
Swallow them up. [*Kneels.*]
Now, by yond marble heaven,
In the due reverence of a sacred vow
I here engage my words.
IAGO: Do not rise yet. *Kneels.*
Witness, you ever-burning lights above!
You elements that clip us round about!
Witness, that here Iago doth give up
The execution of his wit, hands, heart,
To wrong'd Othello's service! Let him command,
And to obey shall be in me remorse,
What bloody business ever. [*They rise.*]
OTHELLO: I greet thy love,
Not with vain thanks, but with acceptance bounteous,

And will upon the instant put thee to 't:
Within these three days let me hear thee say
That Cassio's not alive.

IAGO: My friend is dead; 'tis done at your request;
But let her live.

OTHELLO: Damn her, lewd minx! O, damn her!
Come, go with me apart; I will withdraw
To furnish me with some swift means of death
For the fair devil. Now art thou my lieutenant.

IAGO: I am your own for ever. *Exeunt.*

SCENE FOUR—*Before the Castle.*

Enter DESDEMONA, EMILIA, *and* CLOWN.

DESDEMONA: Do you know, sirrah, where Lieutenant Cassio lies?

CLOWN: I dare not say he lies anywhere.

DESDEMONA: Why, man?

CLOWN: He is a soldier; and for one to say a soldier lies, is stabbing.

DESDEMONA: Go to; where lodges he?

CLOWN: To tell you where he lodges is to tell you where I lie.

DESDEMONA: Can anything be made of this?

CLOWN: I know not where he lodges, and for me to devise a lodging, and say he lies here or he lies there, were to lie in mine own throat.

DESDEMONA: Can you inquire him out, and be edified by report?

CLOWN: I will catechize the world for him; that is, make questions, and by them answer.

DESDEMONA: Seek him, bid him come hither; tell him I have moved my lord in his behalf, and hope all will be well.

CLOWN: To do this is within the compass of man's wit, and therefore I will attempt the doing it. *Exit* CLOWN.

DESDEMONA: Where should I lose that handkerchief, Emilia?
EMILIA: I know not, madam.
DESDEMONA: Believe me, I had rather have lost my purse
Full of crusadoes; and, but my noble Moor
Is true of mind, and made of no such baseness
As jealous creatures are, it were enough
To put him to ill thinking.
EMILIA: Is he not jealous?
DESDEMONA: Who! he? I think the sun where he was born
Drew all such humours from him.
EMILIA: Look! where he comes.
DESDEMONA: I will not leave him now till Cassio
Be call'd to him.

Enter OTHELLO.

How is 't with you, my lord?
OTHELLO: Well, my good lady. [*Aside.*] Oh hardness to dissemble!—
How do you, Desdemona?
DESDEMONA: Well, my good lord.
OTHELLO: Give me your hand. This hand is moist, my lady.
DESDEMONA: It yet has felt no age nor known no sorrow.
OTHELLO: This argues fruitfulness and liberal heart;
Hot, hot, and moist; this hand of yours requires
A sequester from liberty, fasting and prayer,
Much castigation, exercise devout;
For here's a young and sweating devil here,
That commonly rebels. 'Tis a good hand,
A frank one.
DESDEMONA: You may, indeed, say so;
For 'twas that hand that gave away my heart.
OTHELLO: A liberal hand; the hearts of old gave hands,
But our new heraldry is hands not hearts.
DESDEMONA: I cannot speak of this. Come now, your promise.

OTHELLO: What promise, chuck?
DESDEMONA: I have sent to bid Cassio come speak with you.
OTHELLO: I have a salt and sorry rheum offends me.
Lend me thy handkerchief.
DESDEMONA: Here, my lord.
OTHELLO: That which I gave you.
DESDEMONA: I have it not about me.
OTHELLO: Not?
DESDEMONA: No, indeed, my lord.
OTHELLO: That is a fault.
That handkerchief
Did an Egyptian to my mother give;
She was a charmer, and could almost read
The thoughts of people; she told her, while she kept it,
'Twould make her amiable and subdue my father
Entirely to her love, but if she lost it
Or made a gift of it, my father's eye
Should hold her loathed, and his spirits should hunt
After new fancies. She dying gave it me;
And bid me, when my fate would have me wive,
To give it her. I did so: and take heed on 't;
Make it a darling like your precious eye;
To lose 't or give 't away, were such perdition
As nothing else could match.
DESDEMONA: Is 't possible?
OTHELLO: 'Tis true; there's magic in the web of it;
A sibyl, that had number'd in the world
The sun to course two hundred compasses,
In her prophetic fury sew'd the work;
The worms were hallow'd that did breed the silk,
And it was dy'd in mummy which the skilful
Conserv'd of maidens' hearts.
DESDEMONA: Indeed! is 't true?
OTHELLO: Most veritable; therefore look to 't well.

DESDEMONA: Then would to heaven that I had never seen it!
OTHELLO: Ha! wherefore?
DESDEMONA: Why do you speak so startingly and rash?
OTHELLO: Is 't lost? is 't gone? speak, is it out o' the way?
DESDEMONA: Heaven bless us!
OTHELLO: Say you?
DESDEMONA: It is not lost: but what an if it were?
OTHELLO: How!
DESDEMONA: I say, it is not lost.
OTHELLO: Fetch 't, let me see 't.
DESDEMONA: Why, so I can, sir, but I will not now.
This is a trick to put me from my suit:
Pray you let Cassio be receiv'd again.
OTHELLO: Fetch me the handkerchief; my mind misgives.
DESDEMONA: Come, come;
You'll never meet a more sufficient man.
OTHELLO: The handkerchief!
DESDEMONA: I pray, talk me of Cassio.
OTHELLO: The handkerchief!
DESDEMONA: A man that all his time
Hath founded his good fortunes on your love,
Shar'd dangers with you,—
OTHELLO: The handkerchief!
DESDEMONA: In sooth you are to blame.
OTHELLO: Away! *Exit* OTHELLO.
EMILIA: Is not this man jealous?
DESDEMONA: I ne'er saw this before.
Sure, there's some wonder in this handkerchief;
I am most unhappy in the loss of it.
EMILIA: 'Tis not a year or two shows us a man;
They are all but stomachs, and we all but food;
They eat us hungerly, and when they are full
They belch us. Look you! Cassio and my husband.

Enter IAGO *and* CASSIO.

IAGO: There is no other way; 'tis she must do 't:
And, lo! the happiness: go and importune her.
DESDEMONA: How now, good Cassio! what's the news with you?
CASSIO: Madam, my former suit: I do beseech you
That by your virtuous means I may again
Exist, and be a member of his love
Whom I with all the office of my heart
Entirely honour; I would not be delay'd.
If my offence be of such mortal kind
That nor my service past, nor present sorrows,
Nor purpos'd merit in futurity,
Can ransom me into his love again,
But to know so must be my benefit;
So shall I clothe me in a forc'd content,
And shut myself up in some other course
To fortune's alms.
DESDEMONA: Alas! thrice-gentle Cassio!
My advocation is not now in tune;
My lord is not my lord; nor should I know him,
Were he in favour as in humour alter'd.
So help me every spirit sanctified,
As I have spoken for you all my best
And stood within the blank of his displeasure
For my free speech. You must awhile be patient;
What I can do I will, and more I will
Than for myself I dare: let that suffice you.
IAGO: Is my lord angry?
EMILIA: He went hence but now,
And certainly in strange unquietness.
IAGO: Can he be angry? I have seen the cannon,
When it hath blown his ranks into the air,
And, like the devil, from his very arm
Puff'd his own brother; and can he be angry?
Something of moment then; I will go meet him;
There's matter in 't indeed, if he be angry.

DESDEMONA: I prithee, do so. [*Exit* IAGO.] Something, sure, of state,
Either from Venice, or some unhatch'd practice
Made demonstrable here in Cyprus to him,
Hath puddled his clear spirit; and in such cases
Men's natures wrangle with inferior things,
Though great ones are their object. 'Tis even so;
For let our finger ache, and it indues
Our other healthful members ev'n to that sense
Of pain. Nay, we must think men are not gods,
Nor of them look for such observancy
As fits the bridal. Beshrew me much, Emilia,
I was—unhandsome warrior as I am—
Arraigning his unkindness with my soul;
But now I find I had suborn'd the witness,
And he's indicted falsely.

EMILIA: Pray heaven it be state-matters, as you think,
And no conception, nor no jealous toy
Concerning you.

DESDEMONA: Alas the day! I never gave him cause.

EMILIA: But jealous souls will not be answer'd so;
They are not ever jealous for the cause,
But jealous for they are jealous; 'tis a monster
Begot upon itself, born on itself.

DESDEMONA: Heaven keep that monster from Othello's mind!

EMILIA: Lady, amen.

DESDEMONA: I will go seek him. Cassio, walk hereabout;
If I do find him fit, I'll move your suit
And seek to effect it to my uttermost.

CASSIO: I humbly thank your ladyship.

Exeunt DESDEMONA *and* EMILIA.

Enter BIANCA.

BIANCA: 'Save you, friend Cassio!

CASSIO: What make you from home?
How is it with you, my most fair Bianca?

I' faith, sweet love, I was coming to your house.
BIANCA: And I was going to your lodging, Cassio.
What! keep a week away? seven days and nights?
Eight score eight hours? and lovers' absent hours,
More tedious than the dial eight score times?
O weary reckoning!
CASSIO: Pardon me, Bianca,
I have this while with leaden thoughts been press'd,
But I shall, in a more continuate time,
Strike off this score of absence. Sweet Bianca,
[*Giving her* DESDEMONA'*s handkerchief.*]
Take me this work out.
BIANCA: O Cassio! whence came this?
This is some token from a newer friend;
To the felt absence now I feel a cause;
Is 't come to this? Well, well.
CASSIO: Go to, woman!
Throw your vile guesses in the devil's teeth,
From whence you have them. You are jealous now
That this is from some mistress, some remembrance:
No, in good troth, Bianca.
BIANCA: Why, whose is it?
CASSIO: I know not, sweet; I found it in my chamber.
I like the work well; ere it be demanded,—
As like enough it will,—I'd have it copied;
Take it and do 't; and leave me for this time.
BIANCA: Leave you! wherefore?
CASSIO: I do attend here on the general,
And think it no addition nor my wish
To have him see me woman'd.
BIANCA: Why, I pray you?
CASSIO: Not that I love you not.
BIANCA: But that you do not love me.
I pray you, bring me on the way a little
And say if I shall see you soon at night.

CASSIO: 'Tis but a little way that I can bring you,
For I attend here; but I'll see you soon.
BIANCA: 'Tis very good; I must be circumstanc'd. *Exeunt.*

ACT IV

SCENE ONE—*Before the Castle.*

Enter OTHELLO *and* IAGO.

IAGO: Will you think so?
OTHELLO: Think so, Iago!
IAGO: What!
To kiss in private?
OTHELLO: An unauthoriz'd kiss.
IAGO: Or to be naked with her friend in bed
An hour or more, not meaning any harm?
OTHELLO: Naked in bed, Iago, and not mean harm?
It is hypocrisy against the devil:
They that mean virtuously, and yet do so,
The devil their virtue tempts, and they tempt heaven.
IAGO: If they do nothing, 'tis a venial slip;
But if I give my wife a handkerchief,—
OTHELLO: What then?
IAGO: Why, then, 'tis hers, my lord; and, being hers,
She may, I think, bestow 't on any man.
OTHELLO: She is protectress of her honour too;
May she give that?
IAGO: Her honour is an essence that's not seen;
They have it very oft that have it not:
But for the handkerchief,—

OTHELLO: By heaven, I would most gladly have forgot it:—
Thou said'st,—O! it comes o'er my memory,
As doth the raven o'er the infected house,
Boding to all,—he had my handkerchief.
IAGO: Ay, what of that?
OTHELLO: That's not so good now.
IAGO: What,
If I had said I had seen him do you wrong?
Or heard him say,—as knaves be such abroad,
Who having, by their own importunate suit,
Or voluntary dotage of some mistress,
Convinced or supplied them, cannot choose
But they must blab,—
OTHELLO: Hath he said any thing?
IAGO: He hath, my lord; but be you well assur'd,
No more than he'll unswear.
OTHELLO: What hath he said?
IAGO: Faith, that he did—I know not what he did.
OTHELLO: What? what?
IAGO: Lie—
OTHELLO: With her?
IAGO: With her, on her; what you will.

OTHELLO: Lie with her! lie on her! We say, lie on her, when they belie her. Lie with her! that's fulsome. Handkerchief,—confessions,—handkerchief! To confess, and be hanged for his labour. First, to be hanged, and then to confess: I tremble at it. Nature would not invest herself in such shadowing passion without some instruction. It is not words that shake me thus. Pish! Noses, ears, and lips. Is it possible?—Confess!—Handkerchief!—O devil!

Falls in a trance.

IAGO: Work on,
My medicine, work! Thus credulous fools are caught;
And many worthy and chaste dames even thus,
All guiltless, meet reproach. What, ho! my lord!

My lord, I say! Othello!

Enter CASSIO.

How now, Cassio!

CASSIO: What's the matter?

IAGO: My lord is fallen into an epilepsy;
This is his second fit; he had one yesterday.

CASSIO: Rub him about the temples.

IAGO: No, forbear;
The lethargy must have his quiet course,
If not he foams at mouth, and by and by
Breaks out to savage madness. Look! he stirs;
Do you withdraw yourself a little while,
He will recover straight; when he is gone,
I would on great occasion speak with you. [*Exit* CASSIO.]
How is it, general? have you not hurt your head?

OTHELLO: Dost thou mock me?

IAGO: I mock you! no, by heaven.
Would you would bear your fortune like a man!

OTHELLO: A horned man's a monster and a beast.

IAGO: There's many a beast then, in a populous city,
And many a civil monster.

OTHELLO: Did he confess it?

IAGO: Good sir, be a man;
Think every bearded fellow that's but yok'd
May draw with you; there's millions now alive
That nightly lie in those unproper beds
Which they dare swear peculiar; your case is better.
O! 'tis the spite of hell, the fiend's arch-mock,
To lip a wanton in a secure couch,
And to suppose her chaste. No, let me know;
And knowing what I am I know what she shall be.

OTHELLO: O! thou art wise; 'tis certain.

IAGO: Stand you awhile apart;
Confine yourself but in a patient list.
Whilst you were here o'erwhelmed with your grief,—

A passion most unsuiting such a man,—
Cassio came hither; I shifted him away,
And laid good 'scuse upon your ecstasy;
Bade him anon return and here speak with me;
The which he promis'd. Do but encave yourself,
And mark the fleers, the gibes, the notable scorns,
That dwell in every region of his face;
For I will make him tell the tale anew,
Where, how, how oft, how long ago, and when
He hath, and is again to cope your wife:
I say, but mark his gesture. Marry, patience;
Or I shall say you are all in all in spleen,
And nothing of a man.

OTHELLO: Dost thou hear, Iago?
I will be found most cunning in my patience;
But—dost thou hear?—most bloody.

IAGO: That's not amiss;
But yet keep time in all. Will you withdraw?
[OTHELLO *goes apart.*]
Now will I question Cassio of Bianca,
A housewife that by selling her desires
Buys herself bread and clothes; it is a creature
That dotes on Cassio; as 'tis the strumpet's plague
To beguile many and be beguil'd by one.
He, when he hears of her, cannot refrain
From the excess of laughter. Here he comes:

Enter CASSIO.

As he shall smile, Othello shall go mad;
And his unbookish jealousy must construe
Poor Cassio's smiles, gestures, and light behaviour
Quite in the wrong. How do you now, lieutenant?

CASSIO: The worser that you give me the addition
Whose want even kills me.

IAGO: Ply Desdemona well, and you are sure on 't.
[*Speaking lower.*] Now, if this suit lay in Bianca's power,
How quickly should you speed!

CASSIO: Alas! poor caitiff!

OTHELLO: Look! how he laughs already!

IAGO: I never knew woman love man so.

CASSIO: Alas! poor rogue, I think, i' faith, she loves me.

OTHELLO: Now he denies it faintly, and laughs it out.

IAGO: Do you hear, Cassio?

OTHELLO: Now he importunes him
To tell it o'er: go to; well said, well said.

IAGO: She gives it out that you shall marry her;
Do you intend it?

CASSIO: Ha, ha, ha!

OTHELLO: Do you triumph, Roman? do you triumph?

CASSIO: I marry her! what? a customer? I prithee, bear some charity to my wit; do not think it so unwholesome. Ha, ha, ha!

OTHELLO: So, so, so, so. They laugh that win.

IAGO: Faith, the cry goes that you shall marry her.

CASSIO: Prithee, say true.

IAGO: I am a very villain else.

OTHELLO: Have you scored me? Well.

CASSIO: This is the monkey's own giving out: she is persuaded I will marry her, out of her own love and flattery, not out of my promise.

OTHELLO: Iago beckons me; now he begins the story.

CASSIO: She was here even now; she haunts me in every place. I was the other day talking on the sea bank with certain Venetians, and thither comes this bauble, and, by this hand, she falls me thus about my neck,—

OTHELLO: Crying, "O dear Cassio!" as it were; his gesture imports it.

CASSIO: So hangs and lolls and weeps upon me; so hales and pulls me; ha, ha, ha!

OTHELLO: Now he tells how she plucked him to my chamber. O! I see that nose of yours, but not the dog I shall throw it to.

CASSIO: Well, I must leave her company.

IAGO: Before me! look, where she comes.

CASSIO: 'Tis such another fitchew! marry, a perfumed one.

Enter BIANCA.

What do you mean by this haunting of me?

BIANCA: Let the devil and his dam haunt you! What did you mean by that same handkerchief you gave me even now? I was a fine fool to take it. I must take out the work! A likely piece of work, that you should find it in your chamber, and not know who left it there! This is some minx's token, and I must take out the work! There, give it your hobby-horse; wheresoever you had it I'll take out no work on 't.

CASSIO: How now, my sweet Bianca! how now, how now!

OTHELLO: By heaven, that should be my handkerchief!

BIANCA: An you'll come to supper to-night, you may; an you will not, come when you are next prepared for. *Exit.*

IAGO: After her, after her.

CASSIO: Faith, I must; she'll rail in the street else.

IAGO: Will you sup there?

CASSIO: Faith, I intend so.

IAGO: Well, I may chance to see you, for I would very fain speak with you.

CASSIO: Prithee, come; will you?

IAGO: Go to; say no more. [*Exit* CASSIO.]

OTHELLO [*advancing*]: How shall I murder him, Iago?

IAGO: Did you perceive how he laughed at his vice?

OTHELLO: O! Iago!

IAGO: And did you see the handkerchief?

OTHELLO: Was that mine?

IAGO: Yours, by this hand; and to see how he prizes the foolish woman your wife! she gave it him, and he hath given it his whore.

OTHELLO: I would have him nine years a-killing. A fine woman! a fair woman! a sweet woman!

IAGO: Nay, you must forget that.

OTHELLO: Ay, let her rot, and perish, and be damned to-night; for she shall not live. No, my heart is turned to stone; I strike it, and it hurts my hand. O! the world hath not a sweeter creature; she might lie by an emperor's side and command him tasks.

IAGO: Nay, that's not your way.

OTHELLO: Hang her! I do but say what she is. So delicate with her needle! An admirable musician! O, she will sing the savageness out of a bear. Of so high and plenteous wit and invention!

IAGO: She's the worse for all this.

OTHELLO: O! a thousand, a thousand times. And then, of so gentle a condition!

IAGO: Ay, too gentle.

OTHELLO: Nay, that's certain; but yet the pity of it, Iago! O! Iago, the pity of it, Iago!

IAGO: If you are so fond over her iniquity, give her patent to offend; for, if it touch not you, it comes near nobody.

OTHELLO: I will chop her into messes. Cuckold me!

IAGO: O! 'tis foul in her.

OTHELLO: With mine officer!

IAGO: That's fouler.

OTHELLO: Get me some poison, Iago; this night: I'll not expostulate with her, lest her body and beauty unprovide my mind again. This night, Iago.

IAGO: Do it not with poison, strangle her in her bed, even the bed she hath contaminated.

OTHELLO: Good, good; the justice of it pleases; very good.

IAGO: And for Cassio, let me be his undertaker; you shall hear more by midnight.

OTHELLO: Excellent good. [*A trumpet within.*] What trumpet is that same?

IAGO: Something from Venice, sure. 'Tis Lodovico,
Come from the duke; and see, your wife is with him.

Enter LODOVICO, DESDEMONA, *and* Attendants.

LODOVICO: God save you, worthy general!

OTHELLO: With all my heart, sir.

LODOVICO: The duke and senators of Venice greet you.
[*Gives him a letter.*]

OTHELLO: I kiss the instrument of their pleasures.
[*Opens the letter, and reads.*]

DESDEMONA: And what's the news, good cousin Lodovico?

IAGO: I am very glad to see you, signior;
Welcome to Cyprus.

LODOVICO: I thank you. How does Lieutenant Cassio?

IAGO: Lives, sir.

DESDEMONA: Cousin, there's fall'n between him and my lord
An unkind breach; but you shall make all well.

OTHELLO: Are you sure of that?

DESDEMONA: My lord?

OTHELLO [*reads*]: "This fail you not to do, as you will"—

LODOVICO: He did not call; he's busy in the paper.
Is there division 'twixt my lord and Cassio?

DESDEMONA: A most unhappy one; I would do much
To atone them, for the love I bear to Cassio.

OTHELLO: Fire and brimstone!

DESDEMONA: My lord?

OTHELLO: Are you wise?

DESDEMONA: What! is he angry?

LODOVICO: May be the letter mov'd him;
For, as I think, they do command him home,
Deputing Cassio in his government.

DESDEMONA: Trust me, I am glad on 't.
OTHELLO: Indeed!
DESDEMONA: My lord?
OTHELLO: I am glad to see you mad.
DESDEMONA: Why, sweet Othello?
OTHELLO: Devil! [*Strikes her.*]
DESDEMONA: I have not deserved this.
LODOVICO: My lord, this would not be believ'd in Venice,
Though I should swear I saw 't: 'tis very much;
Make her amends, she weeps.
OTHELLO: O devil, devil!
If that the earth could teem with woman's tears,
Each drop she falls would prove a crocodile.
Out of my sight!
DESDEMONA: I will not stay to offend you. [*Going.*]
LODOVICO: Truly, an obedient lady;
I do beseech your lordship, call her back.
OTHELLO: Mistress!
DESDEMONA: My lord?
OTHELLO: What would you with her, sir?
LODOVICO: Who, I, my lord?
OTHELLO: Ay; you did wish that I would make her turn:
Sir, she can turn, and turn, and yet go on,
And turn again; and she can weep, sir, weep;
And she's obedient, as you say, obedient,
Very obedient. Proceed you in your tears.
Concerning this, sir,—O well-painted passion!—
I am commanded home. Get you away;
I'll send for you anon. Sir, I obey the mandate,
And will return to Venice. Hence! avaunt!
[*Exit* DESDEMONA.]
Cassio shall have my place. And, sir, to-night,
I do entreat that we may sup together;
You are welcome, sir, to Cyprus. Goats and monkeys!
Exit.

LODOVICO: Is this the noble Moor whom our full senate
Call all-in-all sufficient? is this the noble nature
Whom passion could not shake? whose solid virtue
The shot of accident nor dart of chance
Could neither graze nor pierce?
IAGO: He is much chang'd.
LODOVICO: Are his wits safe? is he not light of brain?
IAGO: He's that he is; I may not breathe my censure.
What he might be, if, what he might, he is not,
I would to heaven he were!
LODOVICO: What! strike his wife!
IAGO: Faith, that was not so well; yet would I knew
That stroke would prove the worst!
LODOVICO: Is it his use?
Or did the letters work upon his blood,
And new-create this fault?
IAGO: Alas, alas!
It is not honesty in me to speak
What I have seen and known. You shall observe him,
And his courses will denote him so
That I may save my speech. Do but go after
And mark how he continues.
LODOVICO: I am sorry that I am deceiv'd in him. *Exeunt.*

SCENE TWO—*A Room in the Castle.*

Enter OTHELLO *and* EMILIA.

OTHELLO: You have seen nothing, then?
EMILIA: Nor ever heard, nor ever did suspect.
OTHELLO: Yes, you have seen Cassio and her together.
EMILIA: But then I saw no harm, and then I heard
Each syllable that breath made up between them.
OTHELLO: What! did they never whisper?
EMILIA: Never, my lord.

OTHELLO: Nor send you out o' the way?
EMILIA: Never.
OTHELLO: To fetch her fan, her gloves, her mask, nor nothing?
EMILIA: Never, my lord.
OTHELLO: That's strange.
EMILIA: I durst, my lord, to wager she is honest,
Lay down my soul at stake: if you think other,
Remove your thought; it doth abuse your bosom.
If any wretch have put this in your head,
Let heaven requite it with the serpent's curse!
For, if she be not honest, chaste, and true,
There's no man happy; the purest of their wives
Is foul as slander.
OTHELLO: Bid her come hither; go. *Exit* EMILIA.
She says enough; yet she's a simple bawd
That cannot say as much. This is a subtle whore,
A closet lock and key of villainous secrets;
And yet she'll kneel and pray; I have seen her do 't.

Enter DESDEMONA *and* EMILIA.

DESDEMONA: My lord, what is your will?
OTHELLO: Pray, chuck, come hither.
DESDEMONA: What is your pleasure?
OTHELLO: Let me see your eyes;
Look in my face.
DESDEMONA: What horrible fancy's this?
OTHELLO [*to* EMILIA]: Some of your function, mistress;
Leave procreants alone and shut the door;
Cough or cry "hem" if anybody come;
Your mystery, your mystery; nay, dispatch. *Exit* EMILIA.
DESDEMONA: Upon my knees, what doth your speech import?
I understand a fury in your words,
But not the words.
OTHELLO: Why, what art thou?
DESDEMONA: Your wife, my lord; your true
And loyal wife.

OTHELLO: Come, swear it, damn thyself;
Lest, being like one of heaven, the devils themselves
Should fear to seize thee; therefore be double-damn'd;
Swear thou art honest.
DESDEMONA: Heaven doth truly know it.
OTHELLO: Heaven truly knows that thou art false as hell.
DESDEMONA: To whom, my lord? with whom? how am I false?
OTHELLO: Ah! Desdemona; away, away, away!
DESDEMONA: Alas, the heavy day!—Why do you weep?
Am I the motive of these tears, my lord?
If haply you my father do suspect
An instrument of this your calling back,
Lay not your blame on me; if you have lost him,
Why, I have lost him too.
OTHELLO: Had it pleas'd heaven
To try me with affliction, had they rain'd
All kinds of sores, and shames, on my bare head,
Steep'd me in poverty to the very lips,
Given to captivity me and my utmost hopes,
I should have found in some part of my soul
A drop of patience; but, alas! to make me
The fixed figure for the time of scorn
To point his slow unmoving finger at!
Yet could I bear that too; well, very well:
But there, where I have garner'd up my heart,
Where either I must live or bear no life,
The fountain from the which my current runs
Or else dries up; to be discarded thence!
Or keep it as a cistern for foul toads
To knot and gender in! Turn thy complexion there,
Patience, thou young and rose-lipp'd cherubin;
Ay, there, look grim as hell!
DESDEMONA: I hope my noble lord esteems me honest.
OTHELLO: O! Ay; as summer flies are in the shambles,
That quicken even with blowing. O thou weed!
Who art so lovely fair and smell'st so sweet

That the sense aches at thee, would thou hadst ne'er been born.
DESDEMONA: Alas! what ignorant sin have I committed?
OTHELLO: Was this fair paper, this most goodly book,
Made to write "whore" upon? What committed!
Committed! O thou public commoner!
I should make very forges of my cheeks,
That would to cinders burn up modesty,
Did I but speak thy deeds. What committed!
Heaven stops the nose at it and the moon winks,
The bawdy wind that kisses all it meets
Is hush'd within the hollow mine of earth,
And will not hear it. What committed!
Impudent strumpet!
DESDEMONA: By heaven, you do me wrong.
OTHELLO: Are not you a strumpet?
DESDEMONA: No, as I am a Christian.
If to preserve this vessel for my lord
From any other foul unlawful touch
Be not to be a strumpet, I am none.
OTHELLO: What! not a whore?
DESDEMONA: No, as I shall be sav'd.
OTHELLO: Is it possible?
DESDEMONA: O! heaven forgive us.
OTHELLO: I cry you mercy, then;
I took you for that cunning whore of Venice
That married with Othello. You, mistress,
That have the office opposite to Saint Peter,
And keep the gate of hell!

Enter EMILIA.

You, you, ay, you!
We have done our course; there's money for your pains.
I pray you, turn the key and keep our counsel. *Exit.*
EMILIA: Alas! what does this gentleman conceive?
How do you, madam? how do you, my good lady?

DESDEMONA: Faith, half asleep.
EMILIA: Good madam, what's the matter with my lord?
DESDEMONA: With who?
EMILIA: Why, with my lord, madam.
DESDEMONA: Who is thy lord?
EMILIA: He that is yours, sweet lady.
DESDEMONA: I have none; do not talk to me, Emilia;
I cannot weep, nor answer have I none,
But what should go by water. Prithee, to-night
Lay on my bed my wedding sheets: remember:
And call thy husband hither.
EMILIA: Here is a change indeed! *Exit.*
DESDEMONA: 'Tis meet I should be us'd so, very meet.
How have I been behav'd, that he might stick
The small'st opinion on my least misuse?

Enter IAGO *and* EMILIA.

IAGO: What is your pleasure, madam? How is it with you?
DESDEMONA: I cannot tell. Those that do teach young babes
Do it with gentle means and easy tasks;
He might have chid me so; for, in good faith,
I am a child to chiding.
IAGO: What's the matter, lady?
EMILIA: Alas! Iago, my lord hath so bewhor'd her,
Thrown such despite and heavy terms upon her,
As true hearts cannot bear.
DESDEMONA: Am I that name, Iago?
IAGO: What name, fair lady?
DESDEMONA: Such as she says my lord did say I was.
EMILIA: He call'd her whore; a beggar in his drink
Could not have laid such terms upon his callat.[18]
IAGO: Why did he so?

---

[18] [*callat:* basest of low women; harlot.]

DESDEMONA: I do not know; I am sure I am none such.
IAGO: Do not weep, do not weep. Alas the day!
EMILIA: Has she forsook so many noble matches,
  Her father and her country and her friends,
  To be call'd whore? would it not make one weep?
DESDEMONA: It is my wretched fortune.
IAGO: Beshrew him for it!
  How comes this trick upon him?
DESDEMONA: Nay, heaven doth know.
EMILIA: I will be hang'd, if some eternal villain,
  Some busy and insinuating rogue,
  Some cogging, cozening slave, to get some office,
  Have not devis'd this slander; I'll be hang'd else.
IAGO: Fie! there is no such man; it is impossible.
DESDEMONA: If any such there be, heaven pardon him!
EMILIA: A halter pardon him, and hell gnaw his bones!
  Why should he call her whore? who keeps her company?
  What place? what time? what form? what likelihood?
  The Moor's abus'd by some most villainous knave,
  Some base notorious knave, some scurvy fellow.
  O heaven! that such companions thou'dst unfold,
  And put in every honest hand a whip
  To lash the rascals naked through the world,
  Even from the east to the west!
IAGO: Speak within door.
EMILIA: O! fie upon them. Some such squire he was
  That turn'd your wit the seamy side without,
  And made you to suspect me with the Moor.
IAGO: You are a fool; go to.
DESDEMONA: O good Iago,
  What shall I do to win my lord again?
  Good friend, go to him; for, by this light of heaven,
  I know not how I lost him. Here I kneel:
  If e'er my will did trespass 'gainst his love,
  Either in discourse of thought or actual deed,

Or that mine eyes, mine ears, or any sense,
Delighted them in any other form;
Or that I do not yet, and ever did,
And ever will, though he do shake me off
To beggarly divorcement, love him dearly,
Comfort forswear me! Unkindness may do much;
And his unkindness may defeat my life,
But never taint my love. I cannot say "whore":
It does abhor me now I speak the word;
To do the act that might the addition earn
Not the world's mass of vanity could make me.

IAGO: I pray you be content, 'tis but his humour;
The business of the state does him offence,
And he does chide with you.

DESDEMONA: If 'twere no other,—

IAGO: 'Tis but so, I warrant. [*Trumpets within.*]
Hark! how these instruments summon to supper;
The messengers of Venice stay the meat:
Go in, and weep not; all things shall be well.

Exeunt DESDEMONA *and* EMILIA.

Enter RODERIGO.

How now, Roderigo!

RODERIGO: I do not find that thou dealest justly with me.

IAGO: What in the contrary?

RODERIGO: Every day thou daffest me with some device, Iago; and rather, as it seems to me now, keepest from me all conveniency, than suppliest me with the least advantage of hope. I will indeed no longer endure it, nor am I yet persuaded to put up in peace what already I have foolishly suffered.

IAGO: Will you hear me, Roderigo?

RODERIGO: Faith, I have heard too much, for your words and performances are no kin together.

IAGO: You charge me most unjustly.

RODERIGO: With nought but truth. I have wasted myself out of my means. The jewels you have had from me to deliver to Desdemona would half have corrupted a votarist; you have told me she has received them, and returned me expectations and comforts of sudden respect and acquaintance, but I find none.

IAGO: Well; go to; very well.

RODERIGO: Very well! go to! I cannot go to, man; nor 'tis not very well: by this hand, I say, it is very scurvy, and begin to find myself fopped in it.

IAGO: Very well.

RODERIGO: I tell you 'tis not very well. I will make myself known to Desdemona; if she will return me my jewels, I will give over my suit and repent my unlawful solicitation; if not, assure yourself I will seek satisfaction of you.

IAGO: You have said now.

RODERIGO: Ay, and said nothing but what I protest intendment of doing.

IAGO: Why, now I see there's mettle in thee, and even from this instant do build on thee a better opinion than ever before. Give me thy hand, Roderigo; thou hast taken against me a most just exception; but yet, I protest, I have dealt most directly in thy affair.

RODERIGO: It hath not appeared.

IAGO: I grant indeed it hath not appeared, and your suspicion is not without wit and judgment. But, Roderigo, if thou hast that in thee indeed, which I have greater reason to believe now than ever, I mean purpose, courage, and valour, this night show it: if thou the next night following enjoy not Desdemona, take me from this world with treachery and devise engines for my life.

RODERIGO: Well, what is it? is it within reason and compass?

IAGO: Sir, there is especial commission come from Venice to depute Cassio in Othello's place.

RODERIGO: Is that true? why, then Othello and Desdemona return again to Venice.

IAGO: O, no! he goes into Mauritania, and takes away with him the fair Desdemona, unless his abode be lingered here by some accident; wherein none can be so determinate as the removing of Cassio.

RODERIGO: How do you mean, removing of him?

IAGO: Why, by making him uncapable of Othello's place; knocking out his brains.

RODERIGO: And that you would have me do?

IAGO: Ay; if you dare do yourself a profit and a right. He sups to-night with a harlotry, and thither will I go to him; he knows not yet of his honourable fortune. If you will watch his going thence,—which I will fashion to fall out between twelve and one,—you may take him at your pleasure; I will be near to second your attempt, and he shall fall between us. Come, stand not amazed at it, but go along with me; I will show you such a necessity in his death that you shall think yourself bound to put it on him. It is now high supper-time, and the night grows to waste; about it.

RODERIGO: I will hear further reason for this.

IAGO: And you shall be satisfied. *Exeunt.*

SCENE THREE—*Another Room in the Castle.*

Enter OTHELLO, LODOVICO, DESDEMONA, EMILIA, *and* Attendants.

LODOVICO: I do beseech you, sir, trouble yourself no further.

OTHELLO: O! pardon me; 'twill do me good to walk.

LODOVICO: Madam, good night; I humbly thank your ladyship.

DESDEMONA: Your honour is most welcome.

OTHELLO: Will you walk, sir?
O! Desdemona,—

DESDEMONA: My lord?

OTHELLO: Get you to bed on the instant; I will be returned forthwith; dismiss your attendant there; look it be done.

DESDEMONA: I will, my lord.

Exit [OTHELLO, *with* LODOVICO *and* Attendants].

EMILIA: How goes it now? he looks gentler than he did.

DESDEMONA: He says he will return incontinent;[19]
He hath commanded me to go to bed,
And bade me to dismiss you.

EMILIA: Dismiss me!

DESDEMONA: It was his bidding; therefore, good Emilia,
Give me my nightly wearing, and adieu:
We must not now displease him.

EMILIA: I would you had never seen him.

DESDEMONA: So would not I; my love doth so approve him,
That even his stubbornness, his checks and frowns,—
Prithee, unpin me,—have grace and favour in them.

EMILIA: I have laid those sheets you bade me on the bed.

DESDEMONA: All's one. Good faith! how foolish are our minds!
If I do die before thee, prithee, shroud me
In one of those same sheets.

EMILIA: Come, come, you talk.

DESDEMONA: My mother had a maid call'd Barbara;
She was in love, and he she lov'd prov'd mad
And did forsake her; she had a song of "willow";
An old thing 'twas, but it express'd her fortune,
And she died singing it; that song to-night
Will not go from my mind; I have much to do
But to go hang my head all at one side,
And sing it like poor Barbara. Prithee, dispatch.

EMILIA: Shall I go fetch your night-gown?

DESDEMONA: No, unpin me here.
This Lodovico is a proper man.

---

[19] [*incontinent:* immediately.]

EMILIA: A very handsome man.
DESDEMONA: He speaks well.
EMILIA: I know a lady in Venice would have walked barefoot to Palestine for a touch of his nether lip.
DESDEMONA [*singing*]:

"The poor soul sat sighing by a sycamore tree,
Sing all a green willow;
Her hand on her bosom, her head on her knee,
Sing willow, willow, willow:
The frest streams ran by her, and murmur'd her moans;
Sing willow, willow, willow:
Her salt tears fell from her, and soften'd the stones;"—

Lay by these.—
[*Singing.*] "Sing willow, willow, willow"—
Prithee, hie thee; he'll come anon.—
[*Singing.*] "Sing all a green willow must be my garland.
Let nobody blame him, his scorn I approve"—
Nay, that's not next. Hark! who is it that knocks?
EMILIA: It is the wind.
DESDEMONA [*singing*]:

"I call'd my love false love; but what said he then?
Sing willow, willow, willow:
If I court moe women, you'll couch with moe men"—

So, get thee gone; good night. Mine eyes do itch;
Doth that bode weeping?
EMILIA: 'Tis neither here nor there.
DESDEMONA: I have heard it said so. O! these men, these men!
Dost thou in conscience think, tell me, Emilia,
That there be women do abuse their husbands
In such gross kind?
EMILIA: There be some such, no question.
DESDEMONA: Wouldst thou do such a deed for all the world?

EMILIA: Why, would not you?
DESDEMONA: No, by this heavenly light!
EMILIA: Nor I neither by this heavenly light;
I might do 't as well i' the dark.
DESDEMONA: Wouldst thou do such a deed for all the world?
EMILIA: The world is a huge thing; 'tis a great price
For a small vice.
DESDEMONA: In troth, I think thou wouldst not.
EMILIA: In troth, I think I should, and undo 't when I had done. Marry, I would not do such a thing for a joint-ring, nor measures of lawn, nor for gowns, petticoats, nor caps, nor any petty exhibition; but for the whole world, who would not make her husband a cuckold to make him a monarch? I should venture purgatory for 't.
DESDEMONA: Beshrew me, if I would do such a wrong
For the whole world.
EMILIA: Why, the wrong is but a wrong i' the world; and having the world for your labour, 'tis a wrong in your own world, and you might quickly make it right.
DESDEMONA: I do not think there is any such woman.
EMILIA: Yes, a dozen; and as many to the vantage, as would store the world they played for.
But I do think it is their husbands' faults
If wives do fall. Say that they slack their duties,
And pour our treasures into foreign laps,
Or else break out in peevish jealousies,
Throwing restraint upon us; or, say they strike us,
Or scant our former having in despite;
Why, we have galls, and though we have some grace,
Yet have we some revenge. Let husbands know
Their wives have sense like them; they see and smell,
And have their palates both for sweet and sour,
As husbands have. What is it that they do
When they change us for others? Is is sport?
I think it is; and doth affection breed it?
I think it doth; is 't frailty that thus errs?

It is so too; and have not we affections,
Desires for sport, and frailty, as men have?
Then, let them use us well; else let them know,
The ills we do, their ills instruct us so.

DESDEMONA: Good night, good night; heaven me such uses send,
Not to pick bad from bad, but by bad mend! *Exeunt.*

ACT V

SCENE ONE—*A Street near* BIANCA*'s lodging.*

Enter IAGO *and* RODERIGO.

IAGO: Here, stand behind this bulk; straight will he come:
Wear thy good rapier bare, and put it home.
Quick, quick; fear nothing; I'll be at thy elbow.
It makes us, or it mars us; think on that,
And fix most firm thy resolution.

RODERIGO: Be near at hand; I may miscarry in 't.

IAGO: Here, at thy hand: be bold, and take thy stand. [*Retires.*]

RODERIGO: I have no great devotion to the deed;
And yet he has given me satisfying reasons:
'Tis but a man gone: forth, my sword; he dies.

IAGO: I have rubb'd this young quat almost to the sense,
And he grows angry. Now, whether he kill Cassio,
Or Cassio him, or each do kill the other,
Every way makes my gain: live Roderigo,
He calls me to a restitution large
Of gold and jewels that I bobb'd from him,

As gifts to Desdemona;
It must not be: if Cassio do remain,
He hath a daily beauty in his life
That makes me ugly; and, besides, the Moor
May unfold me to him; there stand I in much peril.
No, he must die. But so—I hear him coming.

Enter CASSIO.

RODERIGO: I know his gait; 'tis he. Villain, thou diest!
[*Makes a pass at* CASSIO.]

CASSIO: That thrust had been mine enemy indeed
But that my coat is better than thou know'st;
I will make proof of thine.
[*Draws, and wounds* RODERIGO.]

RODERIGO: O! I am slain. [*Falls.*]
[IAGO *from behind wounds* CASSIO *in the leg, and exit.*]

CASSIO: I am maim'd for ever. Help, ho! murder! murder!
[*Falls.*]

Enter OTHELLO [*at a distance*].

OTHELLO: The voice of Cassio: Iago keeps his word.
RODERIGO: O! villain that I am!
OTHELLO: It is e'en so.
CASSIO: O, help, ho! light! a surgeon!
OTHELLO: 'Tis he: O brave Iago, honest and just,
That hast such noble sense of thy friend's wrong,
Thou teachest me. Minion, your dear lies dead,
And your unblest fate hies; strumpet, I come!
Forth of my heart those charms, thine eyes, are blotted;
Thy bed lust-stain'd shall with lust's blood be spotted.
Exit OTHELLO.

Enter LODOVICO *and* GRATIANO [*at a distance*].

CASSIO: What ho! no watch? no passage? murder! murder!
GRATIANO: 'Tis some mischance; the cry is very direful.
CASSIO: O, help!
LODOVICO: Hark!
RODERIGO: O wretched villain!
LODOVICO: Two or three groan: it is a heavy night;
These may be counterfeits; let's think 't unsafe
To come in to the cry without more help.
RODERIGO: Nobody come? then shall I bleed to death.
LODOVICO: Hark!

Enter IAGO, *with a light.*

GRATIANO: Here's one comes in his shirt, with light and weapons.
IAGO: Who's there? whose noise is this that cries on murder?
LODOVICO: We do not know.
IAGO: Did not you hear a cry?
CASSIO: Here, here! for heaven's sake, help me.
IAGO: What's the matter?
GRATIANO: This is Othello's ancient, as I take it.
LODOVICO: The same indeed; a very valiant fellow.
IAGO: What are you here that cry so grievously?
CASSIO: Iago? O! I am spoil'd, undone by villains!
Give me some help.
IAGO: O me, lieutenant! what villains have done this?
CASSIO: I think that one of them is hereabout,
And cannot make away.
IAGO: O treacherous villains!
[*To* LODOVICO *and* GRATIANO.] What are you there? come in, and give some help.
RODERIGO: O! help me here.
CASSIO: That's one of them.
IAGO: O murderous slave! O villain!
[*Stabs* RODERIGO.]

RODERIGO: O damn'd Iago! O inhuman dog!
IAGO: Kill men i' the dark! Where be these bloody thieves?
How silent is this town! Ho! murder! murder!
What may you be? are you of good or evil?

LODOVICO: As you shall prove us, praise us.
IAGO: Signior Lodovico?
LODOVICO: He, sir.
IAGO: I cry you mercy. Here's Cassio hurt by villains.
GRATIANO: Cassio!
IAGO: How is it, brother?
CASSIO: My leg is cut in two.
IAGO: Marry, heaven forbid!
Light, gentlemen; I'll bind it with my shirt.

Enter BIANCA.

BIANCA: What is the matter, ho? who is 't that cried?
IAGO: Who is 't that cried!
BIANCA: O my dear Cassio! my sweet Cassio!
O Cassio, Cassio, Cassio!
IAGO: O notable strumpet! Cassio, may you suspect
Who they should be that have thus mangled you?
CASSIO: No.
GRATIANO: I am sorry to find you thus; I have been to seek you.
IAGO: Lend me a garter. So. O! for a chair,
To bear him easily hence!
BIANCA: Alas! he faints! O Cassio, Cassio, Cassio!
IAGO: Gentlemen all, I do suspect this trash
To be a party in this injury.
Patience awhile, good Cassio. Come, come.
Lend me a light. Know we this face, or no?
Alas! my friend and my dear countryman,
Roderigo? no: yes, sure, O heaven! Roderigo.
GRATIANO: What! of Venice?
IAGO: Even he, sir: did you know him?
GRATIANO: Know him! Ay.
IAGO: Signior Gratiano? I cry you gentle pardon;
These bloody accidents must excuse my manners,
That so neglected you.
GRATIANO: I am glad to see you.

IAGO: How do you, Cassio? O! a chair, a chair!
GRATIANO: Roderigo! [*A chair brought in.*]
IAGO: He, he, 'tis he.—O! that's well said; the chair:
Some good men bear him carefully from hence;
I'll fetch the general's surgeon. [*To* BIANCA.] For you, mistress,
Save you your labour. He that lies slain here, Cassio,
Was my dear friend. What malice was between you?
CASSIO: None in the world; nor do I know the man.
IAGO [*to* BIANCA]: What! look you pale? O! bear him out o' the air. [CASSIO *and* RODERIGO *are borne off.*]
Stay you, good gentlemen. Look you pale, mistress?—
Do you perceive the gastness of her eye?
Nay, if you stare, we shall hear more anon.
Behold her well; I pray you, look upon her:
Do you see, gentlemen? nay, guiltiness will speak
Though tongues were out of use.

Enter EMILIA.

EMILIA: 'Las! what's the matter? what's the matter, husband?
IAGO: Cassio hath here been set on in the dark
By Roderigo and fellows that are 'scaped:
He's almost slain, and Roderigo dead.
EMILIA: Alas! good gentleman; alas! good Cassio!
IAGO: This is the fruit of whoring. Prithee, Emilia,
Go know of Cassio where he supp'd to-night.
What! do you shake at that?
BIANCA: He supp'd at my house; but I therefore shake not.
IAGO: O! did he so? I charge you, go with me.
EMILIA: Fie, fie upon thee, strumpet!
BIANCA: I am no strumpet, but of life as honest
As you that thus abuse me.
EMILIA: As I! foh! fie upon thee!
IAGO: Kind gentlemen, let's go see poor Cassio dress'd.
Come, mistress, you must tell 's another tale.
Emilia, run you to the citadel,

And tell my lord and lady what hath happ'd.
Will you go on afore? [*Aside.*] This is the night
That either makes me or fordoes me quite. *Exeunt.*

SCENE TWO—*A Bedchamber in the Castle.*

Enter OTHELLO *with a light, and* DESDEMONA *in her bed.*

OTHELLO: It is the cause, it is the cause, my soul;
Let me not name it to you, you chaste stars!
It is the cause. Yet I'll not shed her blood,
Nor scar that whiter skin of hers than snow,
And smooth as monumental alabaster.
Yet she must die, else she'll betray more men.
Put out the light, and then put out the light:
If I quench thee, thou flaming minister,
I can again thy former light restore,
Should I repent me; but once put out thy light,
Thou cunning'st pattern of excelling nature,
I know not where is that Promethean heat
That can thy light relume. When I have pluck'd the rose,
I cannot give it vital growth again,
It needs must wither: I'll smell it on the tree. *Kisses her.*
O balmy breath, that dost almost persuade
Justice to break her sword! One more, one more.
Be thus when thou art dead, and I will kill thee,
And love thee after. One more, and this the last:
So sweet was ne'er so fatal. I must weep,
But they are cruel tears; this sorrow's heavenly,
It strikes where it doth love. She wakes.
DESDEMONA: Who's there? Othello?
OTHELLO: Ay, Desdemona.
DESDEMONA: Will you come to bed, my lord?
OTHELLO: Have you pray'd to-night, Desdemona?

DESDEMONA: Ay, my lord.
OTHELLO: If you bethink yourself of any crime
Unreconcil'd as yet to heaven and grace,
Solicit for it straight.
DESDEMONA: Alas! my lord, what may you mean by that?
OTHELLO: Well, do it, and be brief; I will walk by:
I would not kill thy unprepared spirit;
No; heaven forfend! I would not kill thy soul.
DESDEMONA: Talk you of killing?
OTHELLO: Ay, I do.
DESDEMONA: Then heaven
Have mercy on me!
OTHELLO: Amen! with all my heart.
DESDEMONA: If you say so, I hope you will not kill me.
OTHELLO: Hum!
DESDEMONA: And yet I fear you; for you are fatal then
When your eyes roll so. Why I should fear I know not,
Since guiltiness I know not; but yet I feel I fear.
OTHELLO: Think on thy sins.
DESDEMONA: They are loves I bear to you.
OTHELLO: Ay, and for that thou diest.
DESDEMONA: That death's unnatural that kills for loving.
Alas! why gnaw you so your nether lip?
Some bloody passion shakes your very frame;
These are portents, but yet, I hope, I hope
They do not point on me.
OTHELLO: Peace, and be still!
DESDEMONA: I will so. What's the matter?
OTHELLO: That handkerchief which I so lov'd and gave thee
Thou gav'st to Cassio.
DESDEMONA: No, by my life and soul!
Send for the man and ask him.
OTHELLO: Sweet soul, take heed,
Take heed of perjury; thou'rt on thy death-bed.

DESDEMONA: Ay, but not yet to die.
OTHELLO: Yes, presently;
Therefore confess thee freely of thy sin;
For to deny each article with oath
Cannot remove nor choke the strong conception
That I do groan withal. Thou art to die.
DESDEMONA: Then Lord have mercy on me!
OTHELLO: I say, amen.
DESDEMONA: And have you mercy too! I never did
Offend you in my life; never lov'd Cassio
But with such general warranty of heaven
As I might love; I never gave him token.
OTHELLO: By heaven, I saw my handkerchief in 's hand.
O perjur'd woman! thou dost stone my heart,
And mak'st me call what I intend to do
A murder, which I thought a sacrifice:
I saw the handkerchief.
DESDEMONA: He found it then;
I never gave it him. Send for him hither;
Let him confess a truth.
OTHELLO: He hath confess'd.
DESDEMONA: What, my lord?
OTHELLO: That he hath us'd thee.
DESDEMONA: How? unlawfully?
OTHELLO: Ay.
DESDEMONA: He will not say so.
OTHELLO: No; his mouth is stopp'd:
Honest Iago hath ta'en order for 't.
DESDEMONA: O! my fear interprets. What! is he dead?
OTHELLO: Had all his hairs been lives, my great revenge
Had stomach for them all.
DESDEMONA: Alas! he is betray'd and I undone.
OTHELLO: Out, strumpet! Weep'st thou for him to my face?
DESDEMONA: O! banish me, my lord, but kill me not!

OTHELLO: Down, strumpet!
DESDEMONA: Kill me to-morrow; let me live to-night!
OTHELLO: Nay, if you strive,—
DESDEMONA: But half an hour!
OTHELLO: Being done, there is no pause.
DESDEMONA: But while I say one prayer!
OTHELLO: It is too late.
Smothers her.

EMILIA *at the door.*

EMILIA [*within*]: My lord, my lord! what, ho! my lord, my lord!
OTHELLO: What noise is this? Not dead? not yet quite dead?
I that am cruel am yet merciful;
I would not have thee linger in thy pain.
So, so.
EMILIA [*within*]: What, ho! my lord, my lord!
OTHELLO: Who's there?
EMILIA [*within*]: O! good my lord, I would speak a word with you!
OTHELLO: Yes; 'tis Emilia: by and by. She's dead.
'Tis like she comes to speak of Cassio's death;
The noise was high. Ha! no more moving?
Still as the grave. Shall she come in? Were 't good?
I think she stirs again: no. What's best to do?
If she come in she'll sure speak to my wife:
My wife! my wife! what wife? I have no wife:
O, insupportable! O heavy hour!
Methinks it should be now a huge eclipse
Of sun and moon, and that the affrighted globe
Should yawn at alteration.
EMILIA [*within*]: I do beseech you
That I may speak with you, O good my lord.
OTHELLO: I had forgot thee: O! come in, Emilia:
Soft, by and by: let me the curtains draw.
Where art thou? [*Unlocks the door.*]

Enter EMILIA.

What's the matter with thee now?
EMILIA: O! my good lord, yonder's foul murder done.
OTHELLO: What! now?
EMILIA: But now, my lord.
OTHELLO: It is the very error of the moon;
She comes more near the earth than she was wont,
And makes men mad.
EMILIA: Cassio, my lord, has kill'd a young Venetian
Call'd Roderigo.
OTHELLO: Roderigo kill'd!
And Cassio kill'd?
EMILIA: No, Cassio is not kill'd.
OTHELLO: Not Cassio kill'd! then murder's out of tune,
And sweet revenge grows harsh.
DESDEMONA: O! falsely, falsely murder'd.
EMILIA:: Alas! what cry is that?
OTHELLO: That! what?
EMILIA: Out, and alas! that was my lady's voice:
Help! Help, ho! help! O lady, speak again!
Sweet Desdemona! O! sweet mistress, speak.
DESDEMONA: A guiltless death I die.
EMILIA: O! who hath done this deed?
DESDEMONA: Nobody; I myself; farewell:
Commend me to my kind lord. O! farewell! [*Dies.*]
OTHELLO: Why, how should she be murder'd?
EMILIA: Alas! who knows?
OTHELLO: You heard her say herself it was not I.
EMILIA: She said so; I must needs report the truth.
OTHELLO: She's like a liar gone to burning hell;
Twas I that kill'd her.
EMILIA: O! the more angel she,
And you the blacker devil.
OTHELLO: She turn'd to folly, and she was a whore.
EMILIA: Thou dost belie her, and thou art a devil.

OTHELLO: She was false as water.
EMILIA: Thou art rash as fire to say
That she was false: O! she was heavenly true.
OTHELLO: Cassio did tup her; ask thy husband else.
O! I were damn'd beneath all depth in hell
But that I did proceed upon just grounds
To this extremity. Thy husband knew it all.
EMILIA: My husband!
OTHELLO: Thy husband.
EMILIA: That she was false to wedlock?
OTHELLO: Ay, with Cassio. Nay, had she been true,
If heaven would make me such another world
Of one entire and perfect chrysolite,
I'd not have sold her for it.
EMILIA: My husband!
OTHELLO: Ay, 'twas he that told me first:
An honest man he is, and hates the slime
That sticks on filthy deeds.
EMILIA: My husband!
OTHELLO: What needs this iterance, woman? I say thy husband.
EMILIA: O mistress! villainy hath made mocks with love.
My husband say that she was false!
OTHELLO: He, woman;
I say, thy husband: dost understand the word?
My friend, thy husband, honest, honest Iago.
EMILIA: If he say so, may his pernicious soul
Rot half a grain a day! he lies to the heart:
She was too fond of her most filthy bargain.
OTHELLO: Ha!
EMILIA: Do thy worst:
This deed of thine is no more worthy heaven
Than thou wast worthy her.
OTHELLO: Peace! you were best.
EMILIA: Thou hast not half the power to do me harm
As I have to be hurt. O gull! O dolt!

As ignorant as dirt! thou hast done a deed,—
I care not for thy sword; I'll make thee known,
Though I lost twenty lives. Help! help, ho! help!
The Moor has kill'd my mistress! Murder! murder!

Enter MONTANO, GRATIANO, *and* IAGO [*with* Attendants].

MONTANO: What is the matter? How now, general!
EMILIA: O! are you come, Iago? you have done well,
That men must lay their murders on your neck.
GRATIANO: What is the matter?
EMILIA: Disprove this villain if thou be'st a man:
He says thou told'st him that his wife was false.
I know thou didst not, thou'rt not such a villain.
Speak, for my heart is full.
IAGO: I told him what I thought, and told no more
Than what he found himself was apt and true.
EMILIA: But did you ever tell him she was false?
IAGO: I did.
EMILIA: You told a lie, an odious damned lie;
Upon my soul, a lie, a wicked lie.
She false with Cassio! Did you say with Cassio?
IAGO: With Cassio, mistress. Go to, charm your tongue.
EMILIA: I will not charm my tongue; I am bound to speak.
My mistress here lies murder'd in her bed.
ALL: O heavens forfend!
EMILIA: And your reports have set the murder on.
OTHELLO: Nay, stare not, masters; it is true, indeed.
GRATIANO: 'Tis a strange truth.
MONTANO: O monstrous act!
EMILIA: Villainy! villainy! villainy!
I think upon 't, I think, I smell 't; O villainy!
I thought so then; I'll kill myself for grief.
O! villainy, villainy!
IAGO: What! are you mad? I charge you, get you home.

EMILIA: Good gentlemen, let me have leave to speak:
'Tis proper I obey him, but not now.
Perchance, Iago, I will ne'er go home.
OTHELLO: Oh! Oh! Oh! [OTHELLO *falls on the bed.*]
EMILIA: Nay, lay thee down and roar,
For thou hast kill'd the sweetest innocent
That e'er did lift up eye.
OTHELLO: O! she was foul.
I scarce did know you, uncle. There lies your niece,
Whose breath, indeed, these hands have newly stopp'd:
I know this act shows terrible and grim.
GRATIANO: Poor Desdemona! I am glad thy father's dead.
Thy match was mortal to him, and pure grief
Shore his old thread in twain: did he live now,
This sight would make him do a desperate turn,
Yea, curse his better angel from his side,
And fall to reprobation.
OTHELLO: 'Tis pitiful; but yet Iago knows
That she with Cassio hath the act of shame
A thousand times committed; Cassio confess'd it:
And she did gratify his amorous works
With that recognizance and pledge of love
Which I first gave her. I saw it in his hand:
It was a handkerchief, an antique token
My father gave my mother.
EMILIA: O heaven! O heavenly powers!
IAGO: Come, hold your peace.
EMILIA: 'Twill out, 'twill out; I hold my peace, sir? no;
No, I will speak as liberal as the north;
Let heaven and men and devils, let them all,
All, all, cry shame against me, yet I'll speak.
IAGO: Be wise, and get you home.
EMILIA: I will not.
[IAGO *threatens to stab* EMILIA.]
GRATIANO: Fie!

Your sword upon a woman?
EMILIA: O thou dull Moor! that handkerchief thou speak'st of
I found by fortune and did give my husband;
For often, with a solemn earnestness,
More than, indeed, belong'd to such a trifle,
He begg'd of me to steal it.
IAGO: Villainous whore!
EMILIA: She give it Cassio! no, alas! I found it,
And I did give 't my husband.
IAGO: Filth, thou liest!
EMILIA: By heaven, I do not, I do not, gentlemen.
O murderous coxcomb! what should such a fool
Do with so good a wife?
OTHELLO: Are there no stones in heaven
But what serve for the thunder? Precious villain!
[*He runs at* IAGO. IAGO *stabs* EMILIA, *and exit.*]

GRATIANO: The woman falls: sure, he has kill'd his wife.
EMILIA: Ay, ay; O! lay me by my mistress' side.
GRATIANO: He's gone; but his wife's kill'd.
MONTANO: 'Tis a notorious villain. Take you this weapon,
Which I have here recover'd from the Moor.
Come, guard the door without; let him not pass,
But kill him rather. I'll after that same villain,
For 'tis a damned slave. *Exit* [*with* GRATIANO].
OTHELLO: I am not valiant neither,
But every puny whipster gets my sword.
But why should honour outlive honesty?
Let it go all.
EMILIA: What did thy song bode, lady?
Hark, canst thou hear me? I will play the swan,
And die in music: [*Singing.*]

"Willow, willow, willow."—

Moor, she was chaste; she lov'd thee, cruel Moor;
So come my soul to bliss as I speak true;

So speaking as I think, I die, I die. [*Dies.*]
OTHELLO: I have another weapon in this chamber;
It is a sword of Spain, the ice brook's temper.
O! here it is. Uncle, I must come forth.
GRATIANO [*within*]: If thou attempt it, it will cost thee dear:
Thou hast no weapon, and perforce must suffer.
OTHELLO: Look in upon me then, and speak with me,
Or, naked as I am, I will assault thee.

[*Enter* GRATIANO.]

GRATIANO: What is the matter?
OTHELLO: Behold! I have a weapon;
A better never did itself sustain
Upon a soldier's thigh: I have seen the day,
That with this little arm, and this good sword,
I have made my way through more impediments
Than twenty times your stop: but, O vain boast!
Who can control his fate? 'tis not so now.
Be not afraid, though you do see me weapon'd;
Here is my journey's end, here is my butt,[20]
And very sea-mark of my utmost sail.
Do you go back dismay'd? 'Tis a lost fear;
Man but a rush against Othello's breast,
And he retires. Where should Othello go?
Now how dost thou look now? O ill-starr'd wench!
Pale as thy smock! when we shall meet at compt,[21]
This look of thine will hurl my soul from heaven,
And fiends will snatch at it. Cold, cold, my girl!
Even like thy chastity.
O! cursed, cursed slave. Whip me, ye devils,
From the possession of this heavenly sight!
Blow me about in winds! roast me in sulphur!
Wash me in steep-down gulfs of liquid fire!

---

[20] [*butt:* goal (literally, target).]

[21] [*compt:* reckoning, on the Judgment Day.]

O Desdemona! Desdemona! dead!
Oh! Oh! Oh!

Enter LODOVICO, MONTANO, IAGO [*prisoner*], *and* Officers, CASSIO *in a chair.*

LODOVICO: Where is this rash and most unfortunate man?
OTHELLO: That's he that was Othello; here I am.
LODOVICO: Where is that viper? bring the villain forth.
OTHELLO: I look down towards his feet; but that's a fable.
If that thou be'st a devil, I cannot kill thee.
[*Wounds* IAGO.]

LODOVICO: Wrench his sword from him.
IAGO: I bleed, sir; but not kill'd.
OTHELLO: I am not sorry neither; I'd have thee live;
For, in my sense, 'tis happiness to die.
LODOVICO: O thou Othello! that wert once so good,
Fall'n in the practice of a damned slave,
What shall be said to thee?
OTHELLO: Why, any thing:
An honourable murderer, if you will;
For nought did I in hate, but all in honour.
LODOVICO: This wretch hath part confess'd his villainy:
Did you and he consent in Cassio's death?
OTHELLO: Ay.
CASSIO: Dear general, I never gave you cause.
OTHELLO: I do believe it, and I ask your pardon.
Will you, I pray, demand that demi-devil
Why he hath thus ensnar'd my soul and body?
IAGO: Demand me nothing: what you know, you know:
From this time forth I never will speak word.
LODOVICO: What! not to pray?
GRATIANO: Torments will ope your lips.
OTHELLO: Well, thou dost best.

LODOVICO: Sir, you shall understand what hath befall'n,
Which, as I think, you know not. Here is a letter
Found in the pocket of the slain Roderigo,
And here another; the one of them imports
The death of Cassio to be undertook
By Roderigo.

OTHELLO: O villain!

CASSIO: Most heathenish and most gross!

LODOVICO: Now here's another discontented paper,
Found in his pocket too; and this, it seems,
Roderigo meant to have sent this damned villain,
But that, belike, Iago in the interim
Came in and satisfied him.

OTHELLO: O the pernicious caitiff!
How came you, Cassio, by that handkerchief
That was my wife's?

CASSIO: I found it in my chamber;
And he himself confess'd but even now
That there he dropp'd it for a special purpose
Which wrought to his desire.

OTHELLO: O fool! fool! fool!

CASSIO: There is besides in Roderigo's letter
How he upbraids Iago that he made him
Brave me upon the watch; whereon it came
That I was cast: and even but now he spake,
After long seeming dead, Iago hurt him,
Iago set him on.

LODOVICO: You must forsake this room and go with us;
Your power and your command is taken off,
And Cassio rules in Cyprus. For this slave,
If there be any cunning cruelty
That can torment him much and hold him long,
It shall be his. You shall close prisoner rest,
Till that the nature of your fault be known
To the Venetian state. Come, bring him away.

OTHELLO: Soft you; a word or two before you go.
I have done the state some service, and they know't;
No more of that. I pray you, in your letters,
When you shall these unlucky deeds relate,
Speak of me as I am; nothing extenuate,
Nor set down aught in malice: then must you speak
Of one that lov'd not wisely but too well;
Of one not easily jealous, but, being wrought,
Perplex'd in the extreme; of one whose hand,
Like the base Indian, threw a pearl away
Richer than all his tribe; of one whose subdu'd eyes,
Albeit unused to the melting mood,
Drop tears as fast as the Arabian trees
Their medicinable gum. Set you down this;
And say besides, that in Aleppo once,
Where a malignant and a turban'd Turk
Beat a Venetian and traduc'd the state,
I took by the throat the circumcised dog,
And smote him thus. *Stabs himself.*
LODOVICO: O bloody period!
GRATIANO: All that's spoke is marr'd.
OTHELLO: I kiss'd thee ere I kill'd thee; no way but this,
[*Falling upon* DESDEMONA.]
Killing myself to die upon a kiss. *Dies.*
CASSIO: This did I fear, but thought he had no weapon;
For he was great of heart.
LODOVICO [*to* IAGO]: O Spartan dog!
More fell than anguish, hunger, or the sea,
Look on the tragic loading of this bed;
This is thy work; the object poisons sight;
Let it be hid. Gratiano, keep the house,
And seize upon the fortunes of the Moor,
For they succeed on you. To you, lord governor,
Remains the censure of this hellish villain,
The time, the place, the torture; O! enforce it.
Myself will straight abroad, and to the state
This heavy act with heavy heart relate. *Exeunt.*

DAVID HUME was born in 1711 in Edinburgh, Scotland. His father was a minor laird. Hume attended Edinburgh University when he was twelve and left several years later without receiving a degree. For a time, he studied law, but developed "an insurmountable aversion to everything but the pursuit of philosophy and general learning." His "pursuit" of these brought Hume to a state of nervous collapse in 1729. After recovering, he embarked on a career as a merchant. This palled. In 1734 Hume went to France and spent three years writing his first book, *A Treatise of Human Nature* (1739, 1740). His disappointments with professional reception of his work began with the neglect that the *Treatise* endured. However, Hume continued writing, completing works of philosophy, history, and economics. He worked as a tutor for an insane marquess; was turned down twice for university teaching positions because of his alleged atheism; and served with distinction as secretary to the British embassy in Paris from 1763–65. Hume returned to Edinburgh to live in 1769 and died there in 1776.

From *Hume's Moral and Political Philosophy,* edited by Henry D. Aiken. Publisher: Hafner Press, a division of Macmillan Publishing Co., Inc., 1948. Pages 49–69.

Of Justice and Injustice

SECTION I

JUSTICE, WHETHER A NATURAL OR ARTIFICIAL VIRTUE?

I have already hinted that our sense of every kind of virtue is not natural, but that there are some virtues that produce pleasure and approbation by means of an artifice or contrivance, which arises from the circumstances and necessity of mankind. Of this kind I assert *justice* to be; and shall endeavour to defend this opinion by a short and, I hope, convincing argument, before I examine the nature of the artifice from which the sense of that virtue is derived.

It is evident that, when we praise any actions, we regard only the motives that produced them, and consider the actions as signs or indications of certain principles in the mind and temper. The external performance has no merit. We must look within to find the moral quality. This we cannot do directly; and therefore fix our attention on actions, as on external signs. But these actions are still considered as signs, and the ultimate object of our praise and approbation is the motive that produced them.

After the same manner, when we require any action, or blame a person for not performing it, we always suppose that one in that situation should be influenced by the proper motive of that action, and we esteem it vicious in him to be regardless of it. If we find upon inquiry that the virtuous motive was still powerful over his breast, though checked in its operation by some circumstances unknown to us, we retract our blame and have the same esteem for him as if he had actually performed the action which we require of him.

It appears, therefore, that all virtuous actions derive their merit only from virtuous motives, and are considered merely as signs of those motives. From this principle I conclude that the first virtuous motive which bestows a merit on any action can never be a regard to the virtue of that action, but must be some other natural motive or principle. To suppose that the mere regard to the virtue of the action may be the first motive which produced the action and rendered it virtuous, is to reason in a circle. Before we can have such a regard, the action must be really virtuous; and this virtue must be derived from some virtuous motive; and, consequently, the virtuous motive must be different from the regard to the virtue of the action. A virtuous motive is requisite to render an action virtuous. An action must be virtuous before we can have a regard to its virtue. Some virtuous motive, therefore, must be antecedent to that regard.

Nor is this merely a metaphysical subtilty; but enters into all our reasonings in common life, though, perhaps, we may not be able to place it in such distinct philosophical terms. We blame a father for neglecting his child. Why? Because it shows a want of natural affection which is the duty of every parent. Were not natural affection a duty, the care of children could not be a duty; and it were impossible we could have the duty in our eye in the attention we give to our offspring. In this case, therefore, all men suppose a motive to the action distinct from a sense of duty.

Here is a man that does many benevolent actions: relieves the distressed, comforts the afflicted, and extends his bounty even to the greatest strangers. No character can be more amiable and virtuous. We regard these actions as proofs of the greatest humanity. This humanity bestows a merit on the actions. A regard to this merit is, therefore, a secondary consideration and derived from the antecedent principles of humanity, which is meritorious and laudable.

In short, it may be established as an undoubted maxim that *no action can be virtuous or morally good, unless there be in human nature some motive to produce it distinct from the sense of its morality.*

But may not the sense of morality or duty produce an action without any other motive? I answer, it may; but this is no objection to the present doctrine. When any virtuous motive or principle is common in human nature, a person who feels his heart devoid of that motive may hate himself upon that account, and may perform the action without the motive, from a certain sense of duty, in order to acquire by practice that virtuous principle, or at least to disguise to himself as much as possible his want of it. A man that really feels no gratitude in his temper is still pleased to perform grateful actions, and thinks he has by that means fulfilled his duty. Actions are at first only considered as signs of motives; but it is usual in this case as in all others to fix our attention on the signs, and neglect in some measure the thing signified. But though, on some occasions, a person may perform an action merely out of regard to its moral obligation, yet still this supposes in human nature some distinct principles which are capable of producing the action, and whose moral beauty renders the action meritorious.

Now, to apply all this to the present case, I suppose a person to have lent me a sum of money on condition that it be restored in a few days; and also suppose that after the expiration of the term agreed on he demands the sum; I ask, *What reason or motive have I to restore the money?* It will perhaps be said that my regard to justice, and abhorrence of villainy and knavery, are sufficient reasons for me if I have the least grain of honesty or sense of duty and obligation. And this answer, no doubt, is just and satisfactory to man in his civilized state, and when trained up according to a certain discipline and education. But in his rude and more *natural* condition, if you are pleased to call such a condition natural, this answer would be rejected as perfectly unintelligible and sophistical. For one in that situation would immediately ask you, *Wherein consists this honesty and justice, which you find in restoring a loan and abstaining from the property of others?* It does not surely lie in the external action. It must, therefore, be placed in the motive from which the external action is derived. This motive can never be a regard to the honesty of the action. For it is a plain fallacy to say that a virtuous motive is requisite to render an action honest, and, at the same time, that a regard to the honesty is the motive of the action. We can never have a regard to the virtue of an action, unless the action be antecedently virtuous. No action can be virtuous, but so far as it proceeds from a virtuous motive. A virtuous motive, therefore, must precede the regard to the virtue; and it is impossible that the virtuous motive and the regard to the virtue can be the same.

It is requisite, then, to find some motive to acts of justice and honesty, distinct from our regard to the honesty; and in this lies the great difficulty. For should we say that a concern for our private interest or reputation is the legitimate motive to all honest actions: it would follow that wherever that concern ceases, honesty can no longer have place. But it is certain that self-love, when it acts at its liberty instead of engaging us to honest actions, is the source of all injustice and violence; nor can a man ever correct those vices without correcting and restraining the *natural* movements of that appetite.

But should it be affirmed that the reason or motive of such actions is the *regard to public interest,* to which nothing is more contrary than examples of injustice and dishonesty—should this be said, I would propose the three following considerations as worthy of our attention. First, public interest is not naturally attached to the observation of the rules of justice, but is only connected with it, after an artificial convention for the establishment of these rules, as shall be shown more at large hereafter. Secondly, if we suppose that the loan was secret, and that it is necessary for the interest of the person that the money be restored in the same manner (as when the lender would conceal his riches), in that case the example ceases, and the public is no longer interested in the actions of the borrower, though I suppose there is no moralist who will affirm that the duty and obligation ceases. Thirdly, experience sufficiently proves that men in the ordinary conduct of life look not so far as the public interest, when they pay their creditors, perform their promises, and abstain from theft, and robbery, and injustice of every kind. That is a motive too remote and too sublime to affect the generality of mankind, and operate with any force in actions so contrary to private interest as are frequently those of justice and common honesty.

In general, it may be affirmed that there is no such passion in human minds as the love of mankind, merely as such, independent of personal qualities, of services, or of relation to ourself. It is true, there is no human and indeed no sensible creature whose happiness or misery does not in some measure affect us when brought near us and represented in lively colours; but this proceeds merely from sympathy, and is no proof of such a universal affection to mankind, since this concern extends itself beyond our own species. An affection between the sexes is a passion evidently implanted in human nature; and this passion not only appears in its peculiar symptoms, but also in inflaming every other principle of affection, and raising a stronger love from beauty, wit, kindness, than what would otherwise flow from them. Were there a universal love among all human creatures, it would appear after the same manner. . . .

If public benevolence, therefore, or a regard to the interests of mankind, cannot be the original motive to justice, much less can *private benevolence* or a *regard to the interests of the party concerned* be this motive. For what if he be my enemy and has given me just cause to hate him? What if he be a vicious man and deserves the hatred of all mankind? What if he be a miser and can make no use of what I would deprive him of? What if he be a profligate debauchee and would rather receive harm than benefit from large possessions? What if I be in necessity and have urgent motives to acquire something to my family? In all these cases, the original motive to justice would fail, and, consequently, the justice itself, and along with it all property, right, and obligation. . . .

From all this it follows that we have no real or universal motive for observing the laws of equity but the very equity and merit of that observance; and as no action can be equitable or meritorious, where it cannot arise from some separate motive, there is here an evident sophistry and reasoning in a circle. Unless, therefore, we will allow that nature has established a sophistry, and rendered it necessary and unavoidable, we must allow that the sense of justice and injustice is not derived from nature, but arises artificially, though necessarily, from education and human conventions. . . .

To avoid giving offence, I must here observe that when I deny justice to be a natural virtue, I make use of the word *natural* only as opposed to *artificial*. In another sense of the word, as no principle of the human mind is more natural than a sense of virtue, so no virtue is more natural than justice. Mankind is an inventive species; and where an invention is obvious and absolutely necessary, it may as properly be said to be natural as anything that proceeds immediately from original principles, without the intervention of thought or reflection. Though the rules of justice be *artificial,* they are not *arbitrary*. Nor is the expression improper to call them *laws of nature,* if by *natural* we understand what is common to any species, or even if we confine it to mean what is inseparable from the species.

Section II

Of the Origin of Justice and Property

We now proceed to examine two questions, viz., *concerning the manner in which the rules of justice are established by the artifice of men;* and *concerning the reasons which determine us to attribute to the observance or neglect of these rules a moral beauty and deformity*. These questions will appear afterwards to be distinct. We shall begin with the former.

Of all the animals with which this globe is peopled there is none towards whom nature seems, at first sight, to have exercised more cruelty than towards man, in the numberless wants and necessities with which she has loaded him, and in the slender means which she affords to the relieving of these necessities. In other creatures, these two particulars generally compensate each other. If we consider the lion as a voracious and carnivorous animal, we shall easily discover him to be very necessitous; but if we turn our eye to his make and temper, his agility, his courage, his arms, and his force, we shall find that his advantages hold proportion with his wants. The sheep and ox are deprived of all these advantages; but their appetites are moderate and their food is of easy purchase. In man alone this unnatural conjunction of infirmity and of necessity may be observed in its greatest perfection. Not only the food which is required for his sustenance flies his search and approach, or at least requires his labour to be produced, but he must be possessed of clothes and lodging to defend him against the injuries of the weather; though, to consider him only in himself, he is provided neither with arms, nor force, nor other natural abilities which are in any degree answerable to so many necessities.

It is by society alone he is able to supply his defects, and raise himself up to an equality with his fellow creatures, and even acquire a superiority above them. By society all his infirmities are compensated; and though in that situation his wants multiply every moment upon him, yet his abilities are still more augmented, and leave him in every respect more satisfied and happy than it is possible for him in his savage and solitary condition ever to become. When every individual person labours apart and only for himself, his force is too small to execute any considerable work; his labour being employed in supplying all his different necessities, he never attains a perfection in any particular art; and as his force and success are not at all times equal, the least failure in either of these particulars must be attended with inevitable ruin and misery. Society provides a remedy for these *three* inconveniences. By the conjunction of forces our power is augmented; by the partition of employments our ability increases; and by mutual succour we are less exposed to fortune and accidents. It is by this additional *force, ability,* and *security,* that society becomes advantageous.

But in order to form society, it is requisite not only that it be advantageous, but also that men be sensible of these advantages; and it is impossible in their wild uncultivated state that by study and reflection alone they should ever be able to attain this knowledge. Most fortunately, therefore, there is conjoined to those necessities whose remedies are remote and obscure, another necessity which, having a present and more obvious remedy, may justly be regarded as the first and original principle of human society. This necessity is no other than that natural appetite between the sexes which unites them together and preserves their union till a new tie takes place in their concern for their common offspring. This new concern becomes also a principle of union between the parents and offspring and forms a more numerous society where the parents govern by the advantage of their superior strength and wisdom, and at the same time are restrained in the exercise of their authority by that natural affection which they bear their children. In a little time, custom and habit, operating on the tender minds of the children, makes them sensible of the advantages which they may reap from society, as well as fashions them by degrees for it by rubbing off those rough corners and untoward affections which prevent their coalition.

For it must be confessed that however the circumstances of human nature may render a union necessary, and however those passions of lust and natural affection may seem to render it unavoidable, yet there are other particulars in our *natural temper* and in our *outward circumstances* which are very incommodious, and are even contrary to the requisite conjunction. Among the former we may justly esteem our *selfishness* to be the most considerable. I am sensible that, generally speaking, the representations of this quality have been carried much too far; and that the descriptions which certain philosophers delight so much to form of mankind in this particular are as wide of nature as any accounts of monsters which we meet with in fables and romances. So far from thinking that men have no affection for anything beyond themselves, I am of opinion that, though it be rare to meet with one who loves any single person better than himself, yet it is as rare to meet with one in whom all the kind affections, taken together, do not overbalance all the selfish. Consult common experience; do you not see that, though the whole expense of the family be generally under the direction of the master of it, yet there are few that do not bestow the largest part of their fortunes on the pleasures of their wives and the education of their children, reserving the smallest portion for their own proper use and entertainment? This is what we may observe concerning such as have those endearing ties; and may presume that the case would be the same with others, were they placed in a like situation.

But though this generosity must be acknowledged to the honour of human nature, we may at the same time remark that so noble an affection, instead of fitting men for large societies, is almost as contrary to them as the most narrow selfishness. For while each person loves himself better than any other single person, and in his love to others bears the greatest affection to his relations and acquaintance, this must necessarily produce an opposition of passions, and a consequent opposition of actions which cannot but be dangerous to the new-established union.

It is, however, worth while to remark that this contrariety of passions would be attended with but small danger, did it not concur with a peculiarity in our *outward circumstances* which affords it an opportunity of exerting itself. There are three different species of goods which we are possessed of: the internal satisfaction of our minds, the external advantages of our body, and the enjoyment of such possessions as we have acquired by our industry and good fortune. We are perfectly secure in the enjoyment of the first. The second may be ravished from us, but can be of no advantage to him who deprives us of them. The last only are both exposed to the violence of others, and may be transferred without suffering any loss or alteration; while at the same time there is not a sufficient quantity of them to supply every one's desires and necessities. As the improvement, therefore, of these goods is the chief advantage of society, so the *instability* of their possession, along with their *scarcity,* is the chief impediment.

In vain should we expect to find in *uncultivated nature* a remedy to this inconvenience; or hope for any inartificial principle of the human mind which might control those partial affections, and make us overcome the temptations arising from our circumstances. The idea of justice can never serve to this purpose, or be taken for a natural principle capable of inspiring men with an equitable conduct towards each other. That virtue, as it is now understood, would never have been dreamed of among rude and savage men. For the notion of injury or injustice implies an immorality or vice committed against some other person. And as every immorality is derived from some defect or unsoundness of the passions, and as this defect must be judged of, in a great measure, from the ordinary course of nature in the constitution of the mind, it will be easy to know whether we be guilty of any immorality with regard to others, by considering the natural and usual force of those several affections which are directed towards them. Now, it appears that in the original frame of our mind our strongest attention is confined to ourselves; our next is extended to our relations and acquaintance; and it is only the weakest which reaches to strangers and indifferent persons. This partiality, then, and unequal affection must not only have an influence on our behaviour and conduct in society, but even on our ideas of vice and virtue; so as to make us regard any remarkable transgression of such a degree of partiality, either by too great an enlargement or contraction of the affections, as vicious and immoral. This we may observe in our common judgments concerning actions, where we blame a person who either centers all his affections in his family, or is so regardless of them as, in any opposition of interest, to give the preference to a stranger or mere chance acquaintance. From all which it follows that our natural uncultivated ideas of morality, instead of providing a remedy for the partiality of our affections, do rather conform themselves to that partiality and give it an additional force and influence.

The remedy, then, is not derived from nature but from *artifice;* or, more properly speaking, nature provides a remedy in the judgment and understanding for what is irregular and incommodious in the affections. For when men, from their early education in society, have become sensible of the infinite advantages that result from it, and have besides acquired a new affection to company and conversation, and when they have observed that the principal disturbance in society arises from those goods which we call external, and from their looseness and easy transition from one person to another, they must seek for a remedy by putting these goods as far as possible on the same footing with the fixed and constant advantages of the mind and body. This can be done after no other manner than by a convention entered into by all the members of the society to bestow stability on the possession of those external goods, and leave every one in the peaceable enjoyment of what he may acquire by his fortune and industry. By this means every one knows what he may safely possess; and the passions are restrained in their partial and contradictory motions. Nor is such a restraint contrary to these passions; for if so, it could never be entered into nor maintained; but it is only contrary to their heedless and impetuous movement. Instead of departing from our own interest, or from that of our nearest friends, by abstaining from the possessions of others, we cannot better consult both these interests than by such a convention; because it is by that means we maintain society, which is so necessary to their well-being and subsistence as well as to our own.

This convention is not of the nature of a *promise;* for even promises themselves, as we shall see afterwards, arise from human conventions. It is only a general sense of common interest; which sense all the members of the society express to one another, and which induces them to regulate their conduct by certain rules. I observe that it will be for my interest to leave another in the possession of his goods, *provided* he will act in the same manner with regard to me. He is sensible of a like interest in the regulation of his conduct. When this common sense of interest is mutually expressed and is known to both, it produces a suitable resolution and behaviour. And this may properly enough be called a convention or agreement between us, though without the interposition of a promise; since the actions of each of us have a reference to those of the other, and are performed upon the supposition that something is to be performed on the other part. Two men who pull the oars of a boat do it by an agreement or convention, though they have never given promises to each other. Nor is the rule concerning the stability of possessions the less derived from human conventions, that it arises gradually, and acquires force by a slow progression and by our stability of possessions gradually established

repeated experience of the inconveniences of transgressing it. On the contrary, this experience assures us still more that the sense of interest has become common to all our fellows, and gives us a confidence of the future regularity of their conduct; and it is only on the expectation of this that our moderation and abstinence are founded. In like manner are languages gradually established by human conventions, without any promise. In like manner do gold and silver become the common measures of exchange, and are esteemed sufficient payment for what is of a hundred times their value. . . .

No one can doubt that the convention for the distinction of property and for the stability of possession is of all circumstances the most necessary to the establishment of human society, and that after the agreement for the fixing and observing of this rule there remains little or nothing to be done towards settling a perfect harmony and concord. All the other passions, beside this of interest, are either easily restrained, or are not of such pernicious consequence when indulged. *Vanity* is rather to be esteemed a social passion and a bond of union among men. *Pity* and *love* are to be considered in the same light. And as to *envy* and *revenge,* though pernicious, they operate only by intervals, and are directed against particular persons whom we consider as our superiors or enemies. This avidity alone of acquiring goods and possessions for ourselves and our nearest friends is insatiable, perpetual, universal, and directly destructive of society. There scarce is any one who is not actuated by it; and there is no one who has not reason to fear from it, when it acts without any restraint and gives way to its first and most natural movements. So that, upon the whole, we are to esteem the difficulties in the establishment of society to be greater or less, according to those we encounter in regulating and restraining this passion.

It is certain that no affection of the human mind has both a sufficient force and a proper direction to counterbalance the love of gain, and render men fit members of society by making them abstain from the possessions of others. Benevolence to strangers is too weak for this purpose; and as to the other passions, they rather inflame this avidity, when we observe that the larger our possessions are, the more ability we have of gratifying all our appetites. There is no passion, therefore, capable of controlling the interested affection but the very affection itself, by an alteration of its direction. Now, this alteration must necessarily take place upon the least reflection; since it is evident that the passion is much better satisfied by its restraint than by its liberty, and that, in preserving society, we make much greater advances in the acquiring of possessions than in the solitary and forlorn condition which must follow upon violence and a universal licence. The question, therefore, concerning the wickedness or goodness of human nature enters not in the least into that other question concerning the origin of society; nor is there anything to be considered but the degrees of men's sagacity or folly. For whether the passion of self-interest be esteemed vicious or virtuous, it is all a case, since itself alone restrains it; so that if it be virtuous, men become social by their virtue; if vicious, their vice has the same effect.

Now, as it is by establishing the rule for the stability of possession that this passion restrains itself, if that rule be very abstruse and of difficult invention, society must be esteemed in a manner accidental and the effect of many ages. But if it be found that nothing can be more simple and obvious than that rule; that every parent, in order to preserve peace among his children, must establish it; and that these first rudiments of justice must every day be improved, as the society enlarges—if all this appear evident, as it certainly must, we may conclude that it is utterly impossible for men to remain any considerable time in that savage condition which precedes society, but that his very first state and situation may justly be esteemed social. This, however, hinders not but that philosophers may, if they please, extend their reasoning to the supposed *state of nature;*

provided they allow it to be a mere philosophical fiction, which never had and never could have any reality. Human nature being composed of two principal parts which are requisite in all its actions—the affections and understanding—it is certain that the blind motions of the former, without the direction of the latter, incapacitate men for society; and it may be allowed us to consider separately the effects that result from the separate operations of these two component parts of the mind. The same liberty may be permitted to moral, which is allowed to natural philosophers; and it is very usual with the latter to consider any motion as compounded and consisting of two parts separate from each other, though at the same time they acknowledge it to be in itself uncompounded and inseparable.

This *state of nature,* therefore, is to be regarded as a mere fiction, not unlike that of the *golden age,* which poets have invented; only with this difference, that the former is described as full of war, violence, and injustice, whereas the latter is painted out to us as the most charming and most peaceable condition that can possibly be imagined. The seasons in that first age of nature were so temperate, if we may believe the poets, that there was no necessity for men to provide themselves with clothes and houses as a security against the violence of heat and cold. The rivers flowed with wine and milk, the oaks yielded honey, and nature spontaneously produced her greatest delicacies. Nor were these the chief advantages of that happy age. The storms and tempests were not alone removed from nature; but those more furious tempests were unknown to human breasts, which now cause such uproar and engender such confusion. Avarice, ambition, cruelty, selfishness, were never heard of; cordial affection, compassion, sympathy, were the only movements with which the human mind was yet acquainted. Even the distinction of *mine* and *thine* was banished from that happy race of mortals, and carried with them the very notions of property and obligation, justice and injustice.

This, no doubt, is to be regarded as an idle fiction, but yet deserves our attention because nothing can more evidently show the origin of those virtues which are the subjects of our present inquiry. I have already observed that justice takes its rise from human conventions, and that these are intended as a remedy to some inconveniences which proceed from the concurrence of certain *qualities* of the human mind with the *situation* of external objects. The qualities of the mind are *selfishness* and *limited generosity;* and the situation of external objects is their *easy change,* joined to their *scarcity* in comparison of the wants and desires of men. But however philosophers may have been bewildered in those speculations, poets have been guided more infallibly by a certain taste or common instinct which, in most kinds of reasoning, goes further than any of that art and philosophy with which we have been yet acquainted. They easily perceived, if every man had a tender regard for another, or if nature supplied abundantly all our wants and desires, that the jealousy of interest, which justice supposes, could no longer have place; nor would there be any occasion for those distinctions and limits of property and possession which at present are in use among mankind. Increase to a sufficient degree the benevolence of men, or the bounty of nature, and you render justice useless by supplying its place with much nobler virtues and more valuable blessings. The selfishness of men is animated by the few possessions we have in proportion to our wants; and it is to restrain this selfishness that men have been obliged to separate themselves from the community, and to distinguish between their own goods and those of others.

Nor need we have recourse to the fictions of poets to learn this, but, beside the reason of the thing, may discover the same truth by common experience and observation. It is easy to remark that a cordial affection renders all things common among friends, and that married people in particular mutually lose their property and are unacquainted with the *mine* and *thine,* which are so necessary and yet cause such disturbance in human society. The same effect arises from any alteration in the circumstances of mankind; as when there is such a plenty of anything as satisfies all the desires of men; in which case the distinction of property is entirely lost, and everything remains in common. This we may observe with regard to air and water, though the most valuable of all external objects; and may easily conclude that if men were supplied with everything in the same abundance, or if *every one* had the same affection and tender regard for *every one* as for himself, justice and injustice would be equally unknown among mankind.

Here then is a proposition which, I think, may be regarded as certain, *that it is only from the selfishness and confined generosity of man, along with the scanty provision nature has made for his wants that justice derives its origin.* . . .

Consider that, though the rules of justice are established merely by interest, their connection with interest is somewhat singular, and is different from what may be observed on other occasions. A single act of justice is frequently contrary to *public interest;* and were it to stand alone, without being followed by other acts, may in itself be very prejudicial to society. When a man of merit, of a beneficent disposition, restores a great fortune to a miser or a seditious bigot, he has acted justly and laudably; but the public is a real sufferer. Nor is every single act of justice, considered apart, more conducive to private interest than to public; and it is easily conceived how a man may impoverish himself by a single instance of integrity, and have reason to wish that, with regard to that single act, the laws of justice were for a moment suspended in the universe. But however single acts of justice may be contrary either to public or private interest, it is certain that the whole plan or scheme is highly conducive, or indeed absolutely requisite, both to the support of society and the well-being of every individual. It is impossible to separate the good from the ill. Property must be stable, and must be

selfishness & scarcity of need for justice

fixed by general rules. Though in one instance the public be a sufferer, this momentary ill is amply compensated by the steady prosecution of the rule and by the peace and order which it establishes in society. And even every individual person must find himself a gainer on balancing the account; since without justice society must immediately dissolve, and every one must fall into that savage and solitary condition which is infinitely worse than the worst situation that can possibly be supposed in society. When, therefore, men have had experience enough to observe that, whatever may be the consequence of any single act of justice performed by a single person, yet the whole system of actions concurred in by the whole society is infinitely advantageous to the whole and to every part, it is not long before justice and property take place. Every member of society is sensible of this interest; every one expresses this sense to his fellows along with the resolution he has taken of squaring his actions by it, on condition that others will do the same. No more is requisite to induce any one of them to perform an act of justice, who has the first opportunity. This becomes an example to others; and thus justice establishes itself by a kind of convention or agreement, that is, by a sense of interest, supposed to be common to all, and where every single act is performed in expectation that others are to perform the like. Without such a convention no one would ever have dreamed that there was such a virtue as justice, or have been induced to conform his actions to it. Taking any single act, my justice may be pernicious in every respect; and it is only upon the supposition that others are to imitate my example that I can be induced to embrace that virtue; since nothing but this combination can render justice advantageous, or afford me any motives to conform myself to its rules.

We come now to the *second* question we proposed, viz., *Why we annex the idea of virtue to justice, and of vice to injustice.* This question will not detain us long after the principles which we have already established. All we can say of it at present will be dispatched in a few words; and for further satisfaction the reader must wait till we come to the third part of this book. The *natural* obligation to justice, viz., interest, has been fully explained; but as to the *moral* obligation, or the sentiment of right and wrong, it will first be requisite to examine the natural virtues before we can give a full and satisfactory account of it.

After men have found by experience that their selfishness and confined generosity, acting at their liberty, totally incapacitate them for society, and at the same time have observed that society is necessary to the satisfaction of those very passions, they are naturally induced to lay themselves under the restraint of such rules as may render their commerce more safe and commodious. To the imposition, then, and observance of these rules, both in general and in every particular instance, they are at first induced only by a regard to interest; and this motive, on the first formation of society, is sufficiently strong and forcible. But when society has become numerous and has increased to a tribe or nation, this interest is more remote; nor do men so readily perceive that disorder and confusion follow upon every breach of these rules, as in a more narrow and contracted society. But though in our own actions we may frequently lose sight of that interest which we have in maintaining order, and may follow a lesser and more present interest, we never fail to observe the prejudice we receive either mediately or immediately from the injustice of others, as not being in that case either blinded by passion or biased by any contrary temptation. Nay, when the injustice is so distant from us as no way to affect our interest, it still displeases us, because we consider it as prejudicial to human society and pernicious to every one that approaches the person guilty of it. We partake of their uneasiness by *sympathy;* and as everything which gives uneasiness in human actions, upon the general survey, is called *vice,* and whatever produces satisfaction, in the same manner, is denominated *virtue,* this is the reason why the sense of moral good and evil follows upon justice and injustice. And though this sense in the present case be derived only from contemplating the actions of others, yet we fail not to extend it even to our own actions. The *general rule* reaches beyond those instances from which it arose; while at the same time we naturally *sympathize* with others in the sentiments they entertain of us.

Though this progress of the sentiments be *natural* and even necessary, it is certain that it is here forwarded by the artifice of politicians who, in order to govern men more easily and preserve peace in human society, have endeavoured to produce an esteem for justice and an abhorrence of injustice. This, no doubt, must have its effect; but nothing can be more evident than that the matter has been carried too far by certain writers on morals, who seem to have employed their utmost efforts to extirpate all sense of virtue from among mankind. Any artifice of politicians may assist nature in the producing of those sentiments which she suggests to us, and may even, on some occasions, produce alone an approbation or esteem for any particular action; but it is impossible it should be the sole cause of the distinction we make between vice and virtue; for if nature did not aid us in this particular, it would be in vain for politicians to talk of *honourable* or *dishonourable, praiseworthy* or *blamable*. These words would be perfectly unintelligible and would no more have any idea annexed to them than if they were of a tongue perfectly unknown to us. The utmost politicians can perform is to extend the natural sentiments beyond their original bounds; but still nature must furnish the materials and give us some notion of moral distinctions.

As public praise and blame increase our esteem for justice, so private education and instruction contribute to the same effect. For as parents easily observe that a man is the more useful both to himself and others, the greater degree of probity and honour he is endowed with, and that those principles have greater force when custom and education assist interest and reflection; for these reasons they are induced to inculcate on their children from their earliest infancy the principles of probity, and teach them to regard the observance of those rules by which society is maintained as worthy and honourable, and their violation as base and infamous. By this means the sentiments of honour may take root in their tender minds, and acquire such firmness and solidity that they may fall little short of those principles which are the most essential to our natures, and the most deeply radicated in our internal constitution.

What further contributes to increase their solidity is the interest of our reputation, after the opinion *that a merit or demerit attends justice or injustice* is once firmly established among mankind. There is nothing which touches us more nearly than our reputation, and nothing on which our reputation more depends than our conduct with relation to the property of others. For this reason every one who has any regard to his character, or who intends to live on good terms with mankind, must fix an inviolable law to himself never by any temptation to be induced to violate those principles which are essential to a man of probity and honour.

I shall make only one observation before I leave this subject, viz., that though I assert that in the *state of nature* or that imaginary state which preceded society there be neither justice nor injustice, yet I assert not that it was allowable in such a state to violate the property of others. I only maintain that there was no such thing as property, and, consequently, could be no such thing as justice or injustice.

ALEXIS-CHARLES-HENRI CLÉREL DE TOCQUEVILLE was born in Paris, France in 1805. His great-grandfather had been a statesman, and Tocqueville's father was a prefect who was made a peer of France. Tocqueville's ambitions were political. He served as an apprentice magistrate when a young man, but left France for the United States in 1831 because events of the July Revolution in 1830 had endangered his political fortunes. During his nine-month visit to the U.S., Tocqueville conducted a study of prison reforms with his colleague Gustave de Beaumont. Their book, *On the Penitentiary System in the United States and Its Application in France,* was published in 1833. Another result of the American visit was Tocqueville's own book, *Democracy in America* (1835–40). It established Tocqueville as a political scientist and advanced his career in politics. Tocqueville served in the French Chamber of Deputies, in the Constituent Assembly, and as minister of foreign affairs. But his career had its ups and downs, and Tocqueville was forced to abandon active politics in 1851. He died in Cannes in 1859.

From *Democracy in America,* translated by Henry Reeve. Publisher: Schocken Books, 1961. Portions of Chapters XV and XVI, pages 298–339.

The Power of the Majority

Unlimited Power of the Majority in the United States, and Its Consequences

The very essence of democratic government consists in the absolute sovereignty of the majority; for there is nothing in democratic states which is capable of resisting it. Most of the American constitutions[1] have sought to increase this natural strength of the majority by artificial means.

The legislature is, of all political institutions, the one which is most easily swayed by the wishes of the majority. The Americans determined that the members of the legislature should be elected by the people immediately, and for a very brief term, in order to subject them, not only to the general convictions, but even to the daily passions of their constituents. The members of both Houses are taken from the same class in society, and are nominated in the same manner; so that the modifications of the legislative bodies are almost as rapid and quite as irresistible as those of a single assembly. It is to a legislature thus constituted, that almost all the authority of the government has been entrusted.

But while the law increased the strength of those authorities which of themselves were strong, it enfeebled more and more those which were naturally weak. It deprived the representatives of the executive of all stability and independence; and by

[1] [The constitutions of the states, which Tocqueville considers "in reality the authorities which direct society in America."]

subjecting them completely to the caprices of the legislature, it robbed them of the slender influence which the nature of a democratic government might have allowed them to retain. In several states, the judicial power was also submitted to the elective discretion of the majority; and in all of them its existence was made to depend on the pleasure of the legislative authority, since the representatives were empowered annually to regulate the stipend of the judges.

Custom, however, has done even more than law. A proceeding which will in the end set all the guarantees of representative government at naught, is becoming more and more general in the United States: it frequently happens that the electors, who choose a delegate, point out a certain line of conduct to him, and impose upon him a certain number of positive obligations which he is pledged to fulfil. With the exception of the tumult, this comes to the same thing as if the majority of the populace held its deliberations in the market-place.

Several other circumstances concur in rendering the power of the majority in America not only preponderant, but irresistible. The moral authority of the majority is partly based upon the notion, that there is more intelligence and more wisdom in a great number of men collected together than in a single individual, and that the quantity of legislators is more important than their quality. The theory of equality is in fact applied to the intellect of man; and human pride is thus assailed in its last retreat, by a doctrine which the minority hesitate to admit, and in which they very slowly concur. Like all other powers, and perhaps more than all other powers, the authority of the many requires the sanction of time; at first it enforces obedience by constraint; but its laws are not respected until they have long been maintained.

The right of governing society, which the majority supposes itself to derive from its superior intelligence, was introduced into the United States by the first settlers; and this idea, which would be sufficient of itself to create a free nation, has now been amalgamated with the manners of the people, and the minor incidents of social intercourse.

The French, under the old monarchy, held it for a maxim (which is still a fundamental principle of the English Constitution), that the king could do no wrong; and if he did do wrong, the blame was imputed to his advisers. This notion was highly favourable to habits of obedience; and it enabled the subject to complain of the law, without ceasing to love and honour the lawgiver. The Americans entertain the same opinion with respect to the majority.

The moral power of the majority is founded upon yet another principle, which is, that the interests of the many are to be preferred to those of the few. It will readily be perceived that the respect here professed for the rights of the majority must naturally increase or diminish according to the state of parties. When a nation is divided into several irreconcilable factions, the privilege of the majority is often overlooked, because it is intolerable to comply with its demands.

If there existed in America a class of citizens whom the legislating majority sought to deprive of exclusive privileges, which they had possessed for ages, and to bring down from an elevated station to the level of the ranks of the multitude, it is probable that the minority would be less ready to comply with its laws. But as the United States were colonized by men holding equal rank among themselves, there is as yet no natural or permanent source of dissension between the interests of its different inhabitants.

There are certain communities in which the persons who constitute the minority can never hope to draw over the majority to their side, because they must then give up the very point which is at issue between them. Thus, an aristocracy can never become a majority while it retains its exclusive privileges, and it cannot cede its privileges without ceasing to be an aristocracy.

In the United States, political questions cannot be taken up in so general and absolute a manner; and all parties are willing to recognize the rights of the majority, because they all hope to turn those rights to their own advantage at some future time. The majority therefore in that country exercises a prodigious actual authority, and a moral influence which is scarcely less preponderant; no obstacles exist which can impede, or so much as retard its progress, or which can induce it to heed the complaints of those whom it crushes upon its path. This state of things is fatal in itself and dangerous for the future.

How the Unlimited Power of the Majority Increases in America, the Instability of Legislation, and the Administration Inherent in Democracy

The natural defects of democratic institutions . . . increase at the exact ratio of the power of the majority. To begin with the most evident of them all; the mutability of the laws is an evil inherent in democratic government, because it is natural to democracies to raise men to power in very rapid succession. But this evil is more or less sensible in proportion to the authority and the means of action which the legislature possesses.

In America the authority exercised by the legislative bodies is supreme; nothing prevents them from accomplishing their wishes with celerity, and with irresistible power, while they are supplied by new representatives every year. That is to say, the circumstances which contribute most powerfully to democratic instability, and which admit of the free application of caprice to every object in the State, are here in full operation. In conformity with this principle, America is, at the present day, the country in the world where laws last the shortest time. Almost all the American constitutions have been amended within the course of thirty years: there is therefore not a single American

State which has not modified the principles of its legislation in that lapse of time. As for the laws themselves, a single glance upon the archives of the different states of the Union suffices to convince one, that in America the activity of the legislator never slackens. Not that the American democracy is naturally less stable than any other, but that it is allowed to follow its capricious propensities in the formation of the laws.

The omnipotence of the majority, and the rapid as well as absolute manner in which its decisions are executed in the United States, has not only the effect of rendering the law unstable, but it exercises the same influence upon the execution of the law and the conduct of the public administration. As the majority is the only power which it is important to court, all its projects are taken up with the greatest ardour; but no sooner is its attention distracted, than all this ardour ceases; while in the free states of Europe, the administration is at once independent and secure, so that the projects of the legislature are put into execution, although its immediate attention may be directed to other objects.

In America certain ameliorations are undertaken with much more zeal and activity than elsewhere; in Europe the same ends are promoted by much less social effort, more continuously applied.

Some years ago several pious individuals undertook to ameliorate the condition of the prisons. The public was excited by the statements which they put forward, and the regeneration of criminals became a very popular undertaking. New prisons were built; and, for the first time, the idea of reforming as well as of punishing the delinquent formed a part of prison discipline. But this happy alteration, in which the public had taken so hearty an interest, and which the exertions of the citizens had irresistibly accelerated, could not be completed in a moment. While the new penitentiaries were being erected (and it was the pleasure of the majority that they should be terminated with all possible celerity), the old prisons existed, which still contained a great number of offenders. These gaols became more unwholesome and more corrupt in proportion as the new establishments were beautified and improved, forming a contrast which may readily be understood. The majority was so eagerly employed in founding the new prisons, that those which already existed were forgotten; and as the general attention was diverted to a novel object, the care which had hitherto been bestowed upon the others ceased. The salutary regulations of discipline were first relaxed, and afterwards broken; so that in the immediate neighbourhood of a prison which bore witness to the mild and enlightened spirit of our time, dungeons might be met with which reminded the visitor of the barbarity of the Middle Ages.

Tyranny of the Majority

I hold it to be an impious and an execrable maxim that, politically speaking, a people has a right to do whatsoever it pleases; and yet I have asserted that all authority originates in the will of the majority. Am I, then, in contradiction with myself?

A general law—which bears the name of Justice—has been made and sanctioned, not only by a majority of this or that people, but by a majority of mankind. The rights of every people are consequently confined within the limits of what is just. A nation may be considered in the light of a jury which is empowered to represent society at large, and to apply the great and general law of Justice. Ought such a jury, which represents society, to have more power than the society in which the laws it applies originate?

When I refuse to obey an unjust law, I do not contest the right which the majority has of commanding, but I simply appeal from the sovereignty of the people to the sovereignty of mankind. It has been asserted that a people can never entirely outstep the boundaries of justice and of reason in those affairs which are more peculiarly its own; and that consequently full power may fearlessly be given to the majority by which it is represented. But this language is that of a slave.

A majority taken collectively may be regarded as a being whose opinions, and most frequently whose interests, are opposed to those of another being, which is styled a minority. If it be admitted that a man, possessing absolute power, may misuse that power by wronging his adversaries, why should a majority not be liable to the same reproach? Men are not apt to change their characters by agglomeration; nor does their patience in the presence of obstacles increase with the consciousness of their strength. And for these reasons I can never willingly invest any number of my fellow-creatures with that unlimited authority which I should refuse to any one of them.

I do not think that it is possible to combine several principles in the same government, so as at the same time to maintain freedom, and really to oppose them to one another. The form of government which is usually termed *mixed* has always appeared to me to be a mere chimera. Accurately speaking there is no such thing as a mixed government (with the meaning usually given that word), because in all communities some one principle of action may be discovered, which preponderates over the others. England in the last century, which has been more especially cited as an example of this form of government, was in point of fact an essentially aristocratic state, although it comprised very powerful elements of democracy: for the laws and customs of the country were such, that the aristocracy could not but preponderate in the end, and subject the direction of public affairs to its own will. The error arose from too much attention being paid to the actual struggle which was going on between the nobles and the people, without considering the probable issue of the contest, which was in reality the important point. When a community really has a mixed government, that is to say, when it is equally divided between two adverse principles, it must either pass through a revolution, or fall into complete dissolution.

I am therefore of the opinion that some one social power must always be made to predominate over the others; but I think that liberty is endangered when this power is checked by no obstacles which may retard its course, and force it to moderate its own vehemence.

Unlimited power is in itself a bad and dangerous thing; human beings are not competent to exercise it with discretion; and God alone can be omnipotent, because his wisdom and his justice are always equal to his power. But no power upon earth is so worthy of honour for itself, or of reverential obedience to the rights which it represents, that I would consent to admit its uncontrolled and all-predominant authority. When I see that the right and the means of absolute command are conferred on a people or upon a king, upon an aristocracy or a democracy, a monarchy or a republic, I recognize the germ of tyranny, and I journey onwards to a land of more hopeful institutions.

In my opinion the main evil of the present democratic institutions of the United States does not arise, as is often asserted in Europe, from their weakness, but from their overpowering strength; and I am not so much alarmed at the excessive liberty which reigns in that country, as at the very inadequate securities which exist against tyranny.

When an individual or a party is wronged in the United States, to whom can he apply for redress? If to public opinion, public opinion constitutes the majority; if to the legislature, it represents the majority, and implicitly obeys its injunctions; if to the executive power, it is appointed by the majority and remains a passive tool in its hands; the public troops consist of the majority under arms; the jury is the majority invested with the right of hearing judicial cases; and in certain States even the judges are elected by the majority. However iniquitous or absurd the evil of which you complain may be, you must submit to it as well as you can.

If, on the other hand, a legislative power could be so constituted as to represent the majority without necessarily being the slave of its passions; an executive, so as to retain a certain degree of uncontrolled authority; and a judiciary, so as to remain independent of the two other powers; a government would be formed which would still be democratic without incurring any risk of tyrannical abuse.

I do not say that tyrannical abuses frequently occur in America at the present day; but I maintain that no sure barrier is established against them, and that the causes which mitigate the government are to be found in the circumstances and the manners of the country more than in its laws. . . .

Power Exercised by the Majority in America Upon Opinion

It is in the examination of the display of public opinion in the United States, that we clearly perceive how far the power of the majority surpasses all the powers with which we are acquainted in Europe. Intellectual principles exercise an influence which is so invisible and often so inappreciable, that they baffle the toils of oppression. At the present time the most absolute monarchs in Europe are unable to prevent certain notions, which are opposed to their authority, from circulating in secret throughout their dominions, and even in their courts. Such is not the case in America; as long as the majority is still undecided, discussion is carried on; but as soon as its decision is irrevocably pronounced, a submissive silence is observed; and the friends, as well as the opponents, of the measure, unite in assenting to its propriety. The reason of this is perfectly clear: no monarch is so absolute as to combine all the powers of society in his own hands, and to conquer all opposition, with the energy of a majority, which is invested with the right of making and of executing the laws.

The authority of a king is purely physical, and it controls the actions of the subject without subduing his private will; but the majority possesses a power which is physical and moral at the same time; it acts upon the will as well as upon the actions of men, and it represses not only all contest, but all controversy.

I know of no country in which there is so little true independence of mind and freedom of discussion as in America. In any constitutional state in Europe every sort of religious and political theory may be advocated and propagated abroad; for there is no country in Europe so subdued by any single authority, as not to contain citizens who are ready to protect the man who raises his voice in the cause of truth, from the consequences of his hardihood. If he is unfortunate enough to live under an absolute government, the people is upon his side; if he inhabits a free country, he may find a shelter behind the authority of the throne, if he require one. The aristocratic part of society supports him in some countries, and the democracy in others. But in a nation where democratic institutions exist, organized like those of the United States, there is but one sole authority, one single element of strength and of success, with nothing beyond it.

In America, the majority raises very formidable barriers to the liberty of opinion: within these barriers an author may write whatever he pleases, but he will repent it if he ever step beyond them. Not that he is exposed to the terrors of an auto-da-fé, but he is tormented by the slights and persecutions of daily obloquy. His political career is closed forever, since he has offended the only authority which is able to promote his success. Every sort of compensation, even that of celebrity, is refused to him. Before he published his opinions, he imagined that he held them in common with many others; but no sooner has he declared them openly, than he is loudly censured by his overbearing opponents, while those who think, without having the courage to speak, like him, abandon him in silence. He yields at length, oppressed by the daily efforts he has been making, and he subsides into silence, as if he was tormented by remorse for having spoken the truth.

Fetters and headsmen were the coarse instruments which tyranny formerly employed; but the civilization of our age has refined the arts of despotism, which seemed however to have been sufficiently perfected before. The excesses of monarchical power had devised a variety of physical means of oppression: the democratic republics of the present day have rendered it as entirely an affair of the mind, as that will which it is intended to coerce. Under the absolute sway of an individual despot, the body was attacked in order to subdue the soul; and the soul escaped the blows which were directed against it, and rose superior to the attempt; but such is not the course adopted by tyranny in democratic republics; there the body is left free, and the soul is enslaved. The sovereign can no longer say, "You shall think as I do on pain of death"; but he says, "You are free to think differently from me, and to retain your life, your property, and all that you possess; but if such be your determination, you are henceforth an alien among your people. You may retain your civil rights, but they will be useless to you, for you will never be chosen by your fellow-citizens if you solicit their suffrages; and they will affect to scorn you, if you solicit their esteem. You will remain among men, but you will be deprived of the rights of mankind. Your fellow-creatures will shun you like an impure being; and those who are most persuaded of your innocence will abandon you too, lest they should be shunned in their turn. Go in peace! I have given you your life, but it is an existence incomparably worse than death."

Monarchical institutions have thrown an odium upon despotism; let us beware lest democratic republics should restore oppression, and should render it less odious and less degrading in the eyes of the many, by making it still more onerous to the few.

Works have been published in the proudest nations of the Old World, expressly intended to censure the vices and deride the follies of the times: Labruyère inhabited the palace of Louis XIV when he composed his chapter upon the Great, and Molière criticized the courtiers in the very pieces which were acted before the Court. But the ruling power in the United States is not to be made game of; the smallest reproach irritates its sensibility, and the slightest joke which has any foundation in truth renders it indignant; from the style of its language to the more solid virtues of its character, everything must be made the subject of encomium. No writer, whatever be his eminence, can escape from this tribute of adulation to his fellow-citizens. The majority lives in the perpetual practice of self-applause; and there are certain truths which the Americans can only learn from strangers or from experience.

If great writers have not at present existed in America, the reason is very simply given in these facts; there can be no literary genius without freedom of opinion, and freedom of opinion does not exist in America. The Inquisition has never been able to prevent a vast number of anti-religious books from circulating in Spain. The empire of the majority succeeds much better in the United States, since it actually removes the wish of publishing them. Unbelievers are to be met with in America, but, to say the truth, there is no public organ of infidelity. Attempts have been made by some governments to protect the morality of nations by prohibiting licentious books. In the United States no one is punished for this sort of works, but no one is induced to write them; not because all the citizens are immaculate in their manners, but because the majority of the community is decent and orderly.

In these cases the advantages derived from the exercise of this power are unquestionable; and I am simply discussing the nature of the power itself. This irresistible authority is a constant fact, and its judicious exercise is an accidental occurrence.

Effects of the Tyranny of the Majority Upon the National Character of the Americans

The tendencies which I have just alluded to are as yet very slightly perceptible in political society; but they already begin to exercise an unfavourable influence upon the national character of the Americans. I am inclined to attribute the singular paucity of distinguished political characters to the ever-increasing activity of the despotism of the majority in the United States.

When the American Revolution broke out, they arose in great numbers; for public opinion then served, not to tyrannize over, but to direct the exertions of individuals. Those celebrated men took a full part in the general agitation of mind common at that period, and they attained a high degree of personal fame, which was reflected back upon the nation, but which was by no means borrowed from it.

In absolute governments, the great nobles who are nearest to the throne flatter the passions of the sovereign, and voluntarily truckle to his caprices. But the mass of the nation does not degrade itself by servitude; it often submits from weakness, from habit, or from ignorance, and sometimes from loyalty. Some nations have been known to sacrifice their own desires to those of the sovereign with pleasure and with pride; thus exhibiting a sort of independence in the very act of submission. These peoples are miserable, but they are not degraded. There is a great difference between doing what one does not approve, and feigning to approve what one does; the one is the necessary case of a weak person, the other befits the temper of a lacquey.

In free countries, where everyone is more or less called upon to give his opinion in the affairs of state; in democratic republics, where public life is incessantly commingled with domestic affairs, where the sovereign authority is accessible on every side, and where its attention can almost always be attracted by vociferation, more persons are to be met with who speculate upon its foibles, and live at the cost of its passions, than in absolute monarchies. Not because men are naturally worse in these states than elsewhere, but the temptation is stronger, and of easier access at the same time. The result is a far more extensive debasement of the characters of citizens.

Democratic republics extend the practise of currying favour with the many, and they introduce it into a greater number of classes at once: this is one of the most serious reproaches that can be addressed to them. In democratic states organized on the principles of the American republics, this is more especially the case, where the authority of the majority is so absolute and so irresistible, that a man must give up his rights as a citizen, and almost abjure his quality as a human being, if he intends to stray from the track which it lays down.

In that immense crowd which throngs the avenues to power in the United States, I found very few men who displayed any of that manly candour, and that masculine independence of opinion, which frequently distinguished the Americans in former times, and which constitutes the leading feature in distinguished characters wheresoever they may be found. It seems, at first sight, as if all the minds of the Americans were formed upon one model, so accurately do they correspond in their manner of judging. A stranger does, indeed, sometimes meet with Americans who dissent from these rigourous formularies; with men who deplore the defects of the laws, the mutability and the ignorance of democracy; who even go so far as to observe the evil tendencies which impair the national character, and to point out such remedies as it might be possible to apply; but no one is there to hear these things besides yourself, and you, to whom these secret reflections are confided, are a stranger and a bird of passage. They are very ready to communicate truths which are useless to you, but they continue to hold a different language in public.

If ever these lines are read in America, I am well assured of two things: in the first place, that all who peruse them will raise their voices to condemn me; and in the second place, that very many of them will acquit me at the bottom of their conscience.

I have heard of patriotism in the United States, and it is a virtue which may be found among the people, but never among the leaders of the people. This may be explained by analogy; despotism debases the oppressed much more than the oppressor: in absolute monarchies the king has often great virtues, but the courtiers are invariably servile. It is true that the American courtiers do not say "Sire," or "Your Majesty"—a distinction without a difference. They are forever talking of the natural intelligence of the populace they serve; they do not debate the question as to which of the virtues of their master is pre-eminently worthy of admiration; for they assure him that he possesses all the virtues under heaven without having acquired them, or without caring to acquire them; they do not give him their daughters and their wives to be raised at his pleasure to the rank of his concubines, but, by sacrificing their opinions, they prostitute themselves. Moralists and philosophers in America are not obliged to conceal their opinions under the veil of allegory; but, before they venture upon a harsh truth, they say, "We are aware that the people which we are addressing is too superior to all the weaknesses of human nature to lose the command of its temper for an instant; and we should not hold this language if we were not speaking to men, whom their virtues and their intelligence render more worthy of freedom than all the rest of the world."

It would have been impossible for the sycophants of Louis XIV to flatter more dexterously. For my part, I am persuaded that in all governments, whatever their nature may be, servility will cower to force, and adulation will cling to power. The only means of preventing men from degrading themselves, is to invest no one with that unlimited authority which is the surest method of debasing them.

The Greatest Dangers of the American Republics Proceed from the Unlimited Power of the Majority

Governments usually fall a sacrifice to impotence or to tyranny. In the former case their power escapes from them; it is wrested from their grasp in the latter. Many observers, who have witnessed the anarchy of democratic states, have imagined that the government of those states was naturally weak and impotent. The truth is, that when once hostilities are begun between parties, the government loses its control over society. But I do not think that a democratic power is naturally without force or without resources: say rather, that it is almost always by the abuse of its force, and the misemployment of its resources, that a democratic government fails. Anarchy is almost always produced by its tyranny or its mistakes, but not by its want of strength.

It is important not to confound stability with force, or the greatness of a thing with its duration. In democratic republics, the power which directs society is not stable; for it often changes hands and assumes a new direction. But whichever way it turns, its force is almost irresistible. The governments of the American republics appear to me to be as much centralized as those of the absolute monarchies of Europe, and more energetic than they are. I do not, therefore, imagine that they will perish from weakness.

If ever the free institutions of America are destroyed, that event may be attributed to the unlimited authority of the majority, which may at some future time urge the minorities to desperation, and oblige them to have recourse to physical force.

Anarchy will then be the result, but it will have been brought about by despotism. . . .

Causes Which Mitigate the Tyranny of the Majority in the United States

Absence of Central Administration

I have already pointed out the distinction which is to be made between a centralized government and a centralized administration.[2] The former exists in America, but the latter is nearly unknown there. If the directing power of the American communities had both these instruments of government at its disposal, and united the habit of executing its own commands, to the right of commanding; if, after having established the general principles of government, it descended to the details of public business; and if, having regulated the great interests of the country, it could penetrate into the privacy of individual interests, freedom would soon be banished from the New World.

But in the United States the majority which so frequently displays the tastes and the propensities of a despot, is still destitute of the more perfect instruments of tyranny.

In the American republics the activity of the central government has never as yet been extended beyond a limited number of objects sufficiently prominent to call forth its attention. The secondary affairs of society have never been regulated by its authority; and nothing has hitherto betrayed its desire of interfering in them. The majority is become more and more absolute, but it has not increased the prerogatives of the central government; those great prerogatives have been confined to a certain sphere; and although the despotism of the majority may

---

[2] [Tocqueville observed that the federal government passed laws and issued directives, but that these had to be administered by state and local officials accountable to state and local courts and voters.]

be galling upon one point, it cannot be said to extend to all. However the predominant party in the nation may be carried away by its passions; however ardent it may be in the pursuit of its projects, it cannot oblige all the citizens to comply with its desires in the same manner, and at the same time throughout the country. When the central government which represents that majority has issued a decree, it must entrust the execution of its will to agents, over whom it frequently has no control, and whom it cannot perpetually direct. The townships, municipal bodies, and counties may therefore be looked upon as concealed breakwaters, which check or part the tide of popular excitement. If an oppressive law were passed, the liberties of the people would still be protected by the means by which that law would be put in execution: the majority cannot descend to the details and (as I will venture to style them) the puerilities of administrative tyranny. Nor does the people entertain that full consciousness of its authority, which would prompt it to interfere in these matters; it knows the extent of its natural powers, but it is unacquainted with the increased resources which the art of government might furnish.

This point deserves attention; for if a democratic republic, similar to that of the United States, were ever founded in a country where the power of a single individual had previously subsisted, and the effects of a centralized administration had sunk deep into the habits and the laws of the people, I do not hesitate to assert, that in that country a more insufferable despotism would prevail than any which now exists in the monarchical states of Europe; or indeed than any which could be found on this side of the confines of Asia.

The Profession of the Law in the United States Serves to Counterpoise the Democracy

In visiting the Americans and studying their laws, we perceive that the authority they have entrusted to members of the legal profession, and the influence which these individuals exercise in the government, is the most powerful existing security against the excesses of democracy. . . .

Men who have more especially devoted themselves to legal pursuits, derive from those occupations certain habits of order, a taste for formalities, and a kind of instinctive regard for the regular connection of ideas, which naturally render them very hostile to the revolutionary spirit and the unreflecting passions of the multitude.

The special information which lawyers derive from their studies, ensures them a separate station in society; and they constitute a sort of privileged body in the scale of intelligence. This notion of their superiority perpetually recurs to them in the practice of their profession: they are the masters of a science which is necessary, but which is not very generally known: they serve as arbiters between the citizens; and the habit of directing the blind passions of parties in litigation to their purpose, inspires them with a certain contempt for the judgment of the multitude. To this it may be added, that they naturally constitute *a body;* not by any previous understanding, or by an agreement which directs them to a common end; but the analogy of their studies and the uniformity of their proceedings connect their minds together, as much as a common interest could combine their endeavours.

A portion of the tastes and of the habits of the aristocracy may consequently be discovered in the characters of men in the profession of the law. They participate in the same instinctive love of order and of formalities; and they entertain the same repugnance to the actions of the multitude, and the same secret contempt of the government of the people. I do not mean to say that the natural propensities of lawyers are sufficiently strong to sway them irresistibly; for they, like most other men, are governed by their private interests and the advantages of the moment. . . .

I do not, then, assert that *all* the members of the legal profession are at *all* times the friends of order, and the opponents of innovation, but merely that most of them usually are so. In a community in which lawyers are allowed to occupy, without opposition, that high station which naturally belongs to them, their general spirit will be eminently conservative and anti-democratic. When an aristocracy excludes the leaders of that profession from its ranks, it excites enemies which are the more formidable to its security as they are independent of the nobility by their industrious pursuits; and they feel themselves to be its equal in point of intelligence, although they enjoy less opulence and less power. But whenever an aristocracy consents to impart some of its privileges to these same individuals, the two classes coalesce very readily, and assume, as it were, the consistency of a single order of family interests.

I am, in like manner, inclined to believe, that a monarch will always be able to convert legal practitioners into the most serviceable instruments of his authority. There is a far greater affinity between this class of individuals and the executive power, than there is between them and the people; just as there is a greater natural affinity between the nobles and the monarch, than between the nobles and the people, although the higher orders of society have occasionally resisted the prerogative of the Crown in concert with the lower classes.

Lawyers are attached to public order beyond every other consideration, and the best security of public order is authority. It must not be forgotten, that if they prize the free institutions of their country much, they nevertheless value the legality of those institutions far more: they are less afraid of tyranny than of arbitrary power; and provided that the legislature take upon itself to deprive men of their independence, they are not dissatisfied.

I am therefore convinced that the prince who, in presence of an encroaching democracy, should endeavour to impair the judicial authority in his dominions, and to diminish the political influence of lawyers, would commit a great mistake. He would let slip the substance of authority to grasp at the shadow. He would act more wisely in introducing men connected with the law into the government; and if he entrusted them with the conduct of a despotic power, bearing some marks of violence, that power would most likely assume the external features of justice and of legality in their hands.

The government of democracy is favourable to the political power of lawyers; for when the wealthy, the noble, and the prince are excluded from the government, they are sure to occupy the highest stations, in their own right, as it were, since they are the only men of information and sagacity, beyond the sphere of the people, who can be the object of the popular choice. If, then, they are led by their tastes to combine with the aristocracy and to support the Crown, they are naturally brought into contact with the people by their interests. They like the government of democracy, without participating in its propensities and without imitating its weaknesses; whence they derive a two-fold authority, from it and over it. The people in democratic states does not mistrust the members of the legal profession, because it is well known that they are interested in serving the popular cause; and it listens to them without irritation, because it does not attribute to them any sinister designs. The object of lawyers is not, indeed, to overthrow the institutions of democracy, but they constantly endeavour to give it an impulse which diverts it from its real tendency, by means which are foreign to its nature. Lawyers belong to the people by birth and interest, to the aristocracy by habit and by taste, and they may be looked upon as the natural bond and connecting link of the two great classes of society.

The profession of the law is the only aristocratic element which can be amalgamated without violence with the natural elements of democracy, and which can be advantageously and permanently combined with them. I am not unacquainted with the defects which are inherent in the character of that body of men;

but without this admixture of lawyer-like sobriety with the democratic principle, I question whether democratic institutions could long be maintained; and I cannot believe that a republic could subsist at the present time, if the influence of lawyers in public business did not increase in proportion to the power of the people. . . .

In America there are no nobles or literary men, and the people is apt to mistrust the wealthy; lawyers consequently form the highest political class, and the most cultivated circle of society. They have therefore nothing to gain by innovation, which adds a conservative interest to their natural taste for public order. If I were asked where I place the American aristocracy, I should reply without hesitation, that it is not composed of the rich, who are united together by no common tie, but that it occupies the judicial bench and the bar.

The more we reflect upon all that occurs in the United States, the more shall we be persuaded that the lawyers as a body form the most powerful, if not the only counter-poise to the democratic element. In that country we perceive how eminently the legal profession is qualified by its powers, and even by its defects, to neutralize the vices which are inherent in popular government. When the American people is intoxicated by passion, or carried away by the impetuosity of its ideas, it is checked and stopped by the almost invisible influence of its legal counsellors, who secretly oppose their aristocratic propensities to its democratic instincts, their superstitious attachment to what is antique to its love of novelty, their narrow views to its immense designs, and their habitual procrastination to its ardent impatience.

The courts of justice are the most visible organs by which the legal profession is enabled to control the democracy. The judge is a lawyer, who, independently of the taste for regularity and order which he has contracted in the study of legislation, derives an additional love of stability from his own inalienable functions. His legal attainments have already raised him to a distinguished rank among his fellow-citizens; his political power completes the distinction of his station, and gives him the inclinations natural to privileged classes.

Armed with the power of declaring the laws to be unconstitutional, the American magistrate perpetually interferes in political affairs. He cannot force the people to make laws, but at least he can oblige it not to disobey its own enactments, or to act inconsistently with its own principles. I am aware that a secret tendency to diminish the judicial power exists in the United States; and by most of the constitutions of the several states, the government can, upon the demand of the two Houses of the legislature, remove the judges from their station. By some other constitutions the members of the tribunals are elected, and they are even subjected to frequent re-elections. I venture to predict that these innovations will sooner or later be attended with fatal consequences; and that it will be found out at some future period, that the attack which is made upon the judicial power has affected the democratic republic itself.

It must not, however, be supposed that the legal spirit of which I have been speaking has been confined, in the United States, to the courts of justice; it extends far beyond them. As the lawyers constitute the only enlightened class which the people does not mistrust, they are naturally called upon to occupy most of the public stations. They fill the legislative assemblies, and they conduct the administration; they consequently exercise a powerful influence upon the formation of the law, and upon its execution. The lawyers are, however, obliged to yield to the current of public opinion, which is too strong for them to resist it; but it is easy to find indications of what their conduct would be, if they were free to act as they chose. The Americans, who have made such copious innovations in their political legislation, have introduced very sparing alterations in their civil laws, and that with great difficulty, although those laws are frequently repugnant to their social condition. The reason of this is, that in matters of civil law the majority is obliged to defer to the authority of the legal profession, and that the American lawyers are disinclined to innovate when they are left to their own choice.

It is curious for a Frenchman, accustomed to a very different state of things, to hear the perpetual complaints which are made in the United States, against the stationary propensities of legal men, and their prejudices in favour of existing institutions.

The influence of the legal habits which are common in America extends beyond the limits I have just pointed out. Scarcely any question arises in the United States which does not become, sooner or later, a subject of judicial debate; hence all parties are obliged to borrow the ideas, and even the language usual in judicial proceedings, in their daily controversies. As most public men are, or have been legal practitioners, they introduce the customs and technicalities of their profession into the affairs of the country. The jury extends this habitude to all classes. The language of the law thus becomes, in some measure, a vulgar tongue; the spirit of the law, which is produced in the schools and courts of justice, gradually penetrates beyond their walls into the bosom of society, where it descends to the lowest classes, so that the whole people contracts the habits and the tastes of the magistrate. The lawyers of the United States form a party which is but little feared and scarcely perceived, which has no badge peculiar to itself, which adapts itself with great flexibility to the exigencies of the time, and accommodates itself to all the movements of the social body: but this party extends over the whole community, and it penetrates into all classes of society; it acts upon the country imperceptibly, but it finally fashions it to suit its purposes.

Trial by Jury in the United States Considered as a Political Institution

Since I have been led by my subject to recur to the administration of justice in the United States, I will not pass over this point without adverting to the institution of the jury. Trial by jury may be considered in two separate points of view; as a judicial, and as a political institution. If it entered into my present purpose to inquire, how far trial by jury (more especially in civil cases) contributes to ensure the best administration of justice, I admit that its utility might be contested. As the jury was first introduced at a time when society was in an uncivilized state, and when courts of justice were merely called upon to decide on the evidence of facts, it is not an easy task to adapt it to the wants of a highly civilized community, when the mutual relations of men are multiplied to a surprising extent, and have assumed the enlightened and intellectual character of the age.

My present object is to consider the jury as a political institution; and any other course would divert me from my subject. Of trial by jury, considered as a judicial institution, I shall here say but very few words. When the English adopted trial by jury they were a semi-barbarous people; they are become, in course of time, one of the most enlightened nations of the earth; and their attachment to this institution seems to have increased with their increasing cultivation. They soon spread beyond their insular boundaries to every corner of the habitable globe; some have formed colonies, others independent states; the mother-country has maintained its monarchical constitution; many of its offspring have founded powerful republics; but wherever the English have been, they have boasted of the privilege of trial by jury. They have established it, or hastened to re-establish it, in all their settlements. A judicial institution which obtains the suffrages of a great people for so long a series of ages, which is zealously renewed at every epoch of civilization, in all the climates of the earth, and under every form of human government, cannot be contrary to the spirit of justice.

I turn, however, from this part of the subject. To look upon the jury as a mere judicial institution is to confine our attention to a very narrow view of it; for, however great its influence may be upon the decisions of the law-courts, that influence is very subordinate to the powerful effects which it produces on the destinies of the community at large. The jury is above all a political institution, and it must be regarded in this light in order to be duly appreciated.

By the jury, I mean a certain number of citizens chosen indiscriminately, and invested with a temporary right of judging. Trial by jury, as applied to the repression of crime, appears to me to introduce an eminently republic element into the government, upon the following grounds:

The institution of the jury may be aristocratic or democratic, according to the class of society from which the jurors are selected; but it always preserves its republican character, inasmuch as it places the real direction of society in the hands of the governed, or of a portion of the governed, instead of leaving it under the authority of the government. Force is never more than a transient element of success; and after force comes the notion of right. A government which should only be able to crush its enemies upon a field of battle, would very soon be destroyed. The true sanction of political laws is to be found in penal legislation, and if that sanction be wanting, the law will sooner or later lose its cogency. He who punishes infractions of the law, is therefore the real master of society. Now, the institution of the jury raises the people itself, or at least a class of citizens, to the bench of judicial authority. The institution of the jury consequently invests the people, or that class of citizens, with the direction of society.

In England the jury is returned from the aristocratic portion of the nation; the aristocracy makes the laws, applies the laws, and punishes all infractions of the laws; everything is established upon a consistent footing, and England may with truth be said to constitute an aristocratic republic. In the United States the same system is applied to the whole people. Every American citizen is qualified to be an elector, a juror, and is eligible to office. The system of the jury, as it is understood in America, appears to me to be as direct and as extreme a consequence of the sovereignty of the people, as universal suffrage. These institutions are two instruments of equal power, which contribute to the supremacy of the majority. All the sovereigns who have chosen to govern by their own authority, and to direct society instead of obeying its directions, have destroyed or enfeebled the institution of the jury. The monarchs of the House of Tudor sent to prison jurors who refused to convict, and Napoleon caused them to be returned by his agents. . . .

I am so entirely convinced that the jury is pre-eminently a political institution, that I still consider it in this light when it is applied in civil causes. Laws are always unstable unless they are founded upon the manners of a nation: manners are the only durable and resisting power in a people. When the jury is reserved for criminal offences, the people only witnesses its occasional action in certain particular cases; the ordinary course of life goes on without its interference, and it is considered as an instrument, but not as the only instrument, of obtaining justice. This is true *à fortiori* when the jury is only applied to certain criminal causes.

When, on the contrary, the influence of the jury is extended to civil causes, its application is constantly palpable; it affects all the interests of the community; everyone cooperates in its work: it thus penetrates into all the usages of life, it fashions the human mind to its peculiar forms, and is gradually associated with the idea of justice itself.

The institution of the jury, if confined to criminal causes, is always in danger; but when once it is introduced into civil proceedings, it defies the aggressions of time and of man. If it had been as easy to remove the jury from the manners as from the laws of England, it would have perished under Henry VIII and Elizabeth; and the civil jury did in reality, at that period, save the liberties of the country. In whatever manner the jury be applied, it cannot fail to exercise a powerful influence upon the national character; but this influence is prodigiously increased when it is introduced into civil causes. The jury, and more especially the civil jury, serves to communicate the spirit of the judges to the minds of all the citizens; and this spirit, with the habits which attend it, is the soundest preparation for free institutions. It imbues all classes with a respect for the thing judged, and with the notion of right. If these two elements be removed, the love of independence is reduced to a mere destructive passion. It teaches men to practise equity; every man learns to judge his neighbour as he would himself be judged: and this is especially true of the jury in civil causes; for, while the number of persons who have reason to apprehend a criminal prosecution is small, everyone is liable to have a civil action brought against him. The jury teaches every man not to recoil before the responsibility of his own actions, and impresses him with that manly confidence without which political virtue cannot exist. It invests each citizen with a kind of magistracy; it makes them all feel the duties which they are bound to discharge towards society, and the part which they take in the government. By obliging men to turn their attention to affairs which are not exclusively their own, it rubs off that individual egoism which is the rust of society.

The jury contributes most powerfully to form the judgment, and to increase the natural intelligence of a people; and this is, in my opinion, its greatest advantage. It may be regarded as a gratuitous public school ever open, in which every juror learns to exercise his rights, enters into daily communication with the most learned and enlightened members of the upper classes, and becomes practically acquainted with the laws of his country, which are brought within the reach of his capacity by the efforts of the bar, the advice of the judge, and even by the passions of the parties. I think that the practical intelligence and political good sense of the Americans are mainly attributable to the long use which they have made of the jury in civil causes.

I do not know whether the jury is useful to those who are in litigation; but I am certain it is highly beneficial to those who decide the litigation; and I look upon it as one of the most efficacious means for the education of the people which society can employ.

What I have hitherto said applies to all nations; but the remark I am now about to make is peculiar to the Americans and to democratic peoples. I have already observed that in democracies the members of the legal profession, and the magistrates, constitute the only aristocratic body which can check the irregularities of the people. This aristocracy is invested with no physical power; but it exercises its conservative influence upon the minds of men: and the most abundant source of its authority is the institution of the civil jury. In criminal causes, when society is armed against a single individual, the jury is apt to look upon the judge as the passive instrument of social power, and to mistrust his advice. Moreover, criminal causes are entirely founded upon the evidence of facts which common sense can readily appreciate; upon this ground the judge and the jury are equal. Such, however, is not the case in civil causes; then the judge appears as a disinterested arbiter between the conflicting passions of the parties. The jurors look up to him with confidence, and listen to him with respect, for in this instance their intelligence is completely under the control of his learning. It is the judge who sums up the various arguments with which their memory has been wearied out, and who guides them through the devious course of the proceedings; he points their attention to the exact question of fact, which they are called upon to solve, and he puts the answer to the question of law into their mouths. His influence upon their verdict is almost unlimited.

If I am called upon to explain why I am but little moved by the arguments derived from the ignorance of jurors in civil causes, I reply, that in these proceedings, whenever the question to be solved is not a mere question of fact, the jury has only the semblance of a judicial body. The jury sanctions the decision of the judge; they, by the authority of society which they represent, and he, by that of reason and of law.

In England and in America the judges exercise an influence upon criminal trials which the French judges have never possessed. The reason of this difference may easily be discovered; the English and American magistrates establish their authority in civil causes, and only transfer it afterwards to tribunals of another kind, where that authority was not acquired. In some cases (and they are frequently the most important ones), the American judges have the right of deciding causes alone. Upon these occasions they are, accidentally, placed in the position which the French judges habitually occupy: but they are invested with far more power than the latter; they are still surrounded by the reminiscence of the jury, and their judgment has almost as much authority as the voice of the community at large, represented by that institution. Their influence extends beyond the limits of the courts; in the recreations of private life as well as in the turmoil of public business, abroad and in the legislative assemblies, the American judge is constantly surrounded by men who are accustomed to regard his intelligence as superior to their own; and after having exercised his power in the decision of causes, he continues to influence the habits of thought, and the characters of the individuals who took a part in his judgment.

The jury, then, which seems to restrict the rights of magistracy, does in reality consolidate its power; and in no country are the judges so powerful as there, where the people partakes their privileges. It is more especially by means of the jury in civil causes that the American magistrates imbue all classes of society with the spirit of their profession. Thus the jury, which is the most energetic means of making the people rule, is also the most efficacious means of teaching it to rule well.

GEORG SIMMEL was born in 1858 in Berlin, Germany. His father was a partner in a chocolate factory and died when Simmel was a boy. A guardian later left Simmel an inheritance, freeing him to pursue a scholarly career. Simmel studied history and philosophy at the University of Berlin, earning his doctorate in philosophy in 1881. He was appointed a lecturer and professor at Berlin from 1885 to 1900; from 1900 to 1914, Simmel taught at the University of Strasbourg. Simmel was a pioneer in the field of sociology and its methodology. He also wrote and lectured on philosophy. *The Problems of the Philosophy of History* (1892) was his first book. In *The Philosophy of Money* (1900), Simmel applied his sociological theories to economics. *The Sociology of Georg Simmel,* his collected work in translation, was published in 1950. Simmel died of cancer in 1918.

From *The Philosophy of Money,* translated by Tom Bottomore and David Frisby. Publisher: Routledge and Kegan Paul, Ltd., 1978. Pages 283–303.

Individual Freedom

Freedom exists in conjunction with duties

The development of each human fate can be represented as an uninterrupted alternation between bondage and release, obligation and freedom. This initial appraisal, however, presents us with a distinction whose abruptness is tempered by closer investigation. For what we regard as freedom is often in fact only a change of obligations; as a new obligation replaces one that we have borne hitherto, we sense above all that the old burden has been removed. Because we are free from it, we seem at first to be completely free—until the new duty, which initially we bear, as it were, with hitherto untaxed and therefore particularly strong sets of muscles, makes its weight felt as these muscles, too, gradually tire. The process of liberation now starts again with this new duty, just as it had ended at this very point. This pattern is not repeated in a quantitatively uniform manner in all forms of bondage. Rather, there are some with which the note of freedom is associated longer, more intensively, and more consciously than with others. Some accomplishments that are no less rigidly required of some than of others and that are generally no less demanding on the powers of the personality nonetheless seem to allow the personality a particularly large amount of freedom. The difference in obligations which leads to this difference in the freedom compatible with obligations is of the following type. Each obligation that does not exist with regard to a mere idea corresponds to the right of someone else to make demands. For this reason, moral philosophy always identifies ethical freedom with those *obligations* imposed by an ideal or social imperative or by one's own ego. The other person's demands can consist of the personal actions and deeds of the person under obligation. Or they can be realized at least in the immediate outcome of personal labor. Or, finally, it need only be a certain object, the use of which someone can rightly lay claim to, although he has no influence whatsoever concerning the manner in which the person under obligation procures this object for him. This scale is also that of the degrees of freedom that exist with the performance of a duty.

The graduations of this freedom depend on whether the duties are directly personal or apply only to the products of labour

Naturally, every obligation is generally resolved through the personal actions of the human subject, but it makes a great deal of difference as to whether the rights of the person entitled to some service extend directly to the person under obligation himself or simply to the product of the latter's labour or, finally, to the product in itself—regardless of whether the person under obligation acquired the product through his own labor or not. Even if the advantages of the entitled person remained objectively the same, the first of these forms of obligation would completely bind the obligated person, the second would permit him a little more latitude and the third considerable latitude. The most extreme example of the first type is slavery; in this case, the obligation does not involve a service that is in some way objectively defined, but instead refers to the person himself who performs the service. It includes the employment of all the available energy of the human subject. If, under modern conditions, duties that involve the whole capacity to work as such but not the objectively defined result of this capacity—as with certain categories of workers, civil servants, and domestic servants—do not offend against freedom in too crass a manner, then this is a result either of the temporal restriction in the periods of service or of the possibility of selecting the people whom one wishes to be obligated to, or a result of the magnitude of what is offered in return, which makes the obligated person feel, at the same time, that he too has rights. The bondsmen are about at this level, as long as they belong completely, and with their entire working capacity, to their lord's domain, or rather, as long as their services are "unmeasured."

The transition to the second level occurs when the services are temporally limited (but this does not imply that this level was always later historically; on the contrary, the deterioration in peasants' freedom very often leads from the second to the first level of obligation). This second level is definitely reached when, instead of a fixed amount of labor time and energy, a specific product of labour is required. At this level, one can observe a certain gradation, namely, that the manorial serf had to hand over either an aliquot part of the yield from the soil—for example, every tenth sheaf of corn—or a permanently fixed amount of corn, cattle, honey, etc. Although the latter arrangement might possibly be the more severe and more difficult, it nonetheless creates great individual freedom for the obligated person, for it makes the lord of the manor more indifferent towards the peasant's type of husbandry. If the serf only produces what is sufficient for his payment to the lord of the manor, then the latter has no interest in the total yield. . . . One finds a similar phenomenon today when talented people, who are forced to work for a wage, prefer to work for a company with its strictly objective organization rather than for an individual employer; or when a shortage of domestic servants occurs because girls prefer factory work to service with people of authority, where they are certainly in a better position materially but feel themselves less free in their subordination to individual personalities.

Money payment as the form most congruent with personal freedom

The third level, where the person is actually excluded from the product and the demands no longer extend to him, is reached with the replacement of payment in kind by money payment. For this reason, it has been regarded, to some extent, as a *magna charta* of personal freedom in the domain of civil law. Classical Roman law declared that, if a payment in kind were refused, then any demand for payment could be met with money. This is, therefore, the right to buy oneself out of a personal obligation by means of money. The lord of the manor who can demand a quantity of beer or poultry or honey from a serf thereby determines the activity of the latter in a certain direction. But the moment he imposes merely a money levy the peasant is free, insofar as he can decide whether to keep bees or cattle or anything else. Formally, in the sphere of personal labour services, the same process takes place with the right of appointing another person as a substitute, and the other party has to accept the latter unless his competence is in doubt. This right, which sets the whole conception of the relationship on a new basis, must often be fought for since it is felt that, like the right to make payment in money, it is a step on the way towards a dissolution of the entire obligation. . . .

The greatest step forward in the process of liberation is achieved through a development within money payment itself when a single capital payment replaces the periodical levy. Even if the objective value is identical in both forms of payment, the effect upon the human subject is quite different. As we have pointed out, the various payments levied certainly give the obligated person complete freedom in terms of his actions, provided that he obtains the money required. But the regularity of the payments forces such action into a fixed scheme, imposed by an alien power, and so it is only with the capitalization of the payments that the form of all kinds of obligations is attained which corresponds to the greatest personal freedom.

Thus, it is only with the capital payment that the obligation is entirely converted into a money payment, whereas the money levy with its regular recurrence still preserves at least a formal element of bondage over and above the value required in payments. This distinction is manifested in the following manner. In the thirteenth century and later the English Parliament often decided that the counties had to provide a certain number of soldiers or workers for the king. The representatives of the counties, however, regularly replaced the provision of men with a money payment. But no matter how much personal freedom was saved in this manner, there is a fundamental distinction between this and the rights and freedoms that the English people purchased from their monarchs through single votes on money. If the person receiving the capital is then freed from all the insecurities to which he is subjected in the case of individual levies, then the corresponding equivalent on the side of the obligated person is that his freedom is converted from the unstable form that it possesses when recurrent payments have to be made into a stable form. The freedom of the English people with regard to their monarchs depends partially upon the fact that, by means of capital payments, the people had settled matters with their king once and for all with respect to certain rights: a document from Henry III states, for example, "*pro hac concessione dederunt nobis quintam decimam partem omnium movilium suorum.*"[1] It is not in spite of but precisely because of the fact that such an agreement concerning the freedoms of the people reveals a somewhat brutal, external, and mechanical character that it implies the most complete antagonism contrary to the feelings of the king that "no piece of paper should come

[1] ["In return for this concession, they gave us the fifteenth part of all their moveable property."]

between him and his people." Yet precisely for this reason it also constitutes a radical abolition of all the imponderables of more emotional relations which, when freedoms are attained in a form less tied to money transactions, often provide the means for revoking them or making them illusory. . . .

In such cases of the replacement of natural services by money payments the advantage is usually mutual. This is a most remarkable fact which calls for an analysis within a wider context. If one starts out from the assumption that the quantity of goods available for consumption is limited; that this quantity does not satisfy the given demands; that, finally, "the world has been given away," that is, that in general every good has its owner, then it follows that whatever is given to one person must be taken away from another. Even if one disregards all cases where this obviously does not hold, there still remain countless others where the satisfaction of one person's needs is at the expense of that of another. If one were to consider this as the, or as one, characteristic or basis of our economic life, then it would accord with all those world views that hold as immutable the total amount of values given to mankind—such as ethics, happiness, knowledge—so that only the forms and agents of these values can change. Schopenhauer is inclined to assume that the amount of suffering and joy that each individual experiences is predetermined by his essential nature, that this amount can neither be exceeded nor remain void, and that all extraneous circumstances to which we are accustomed to ascribe our situation only represent a difference in the form in which we experience that unchangeable amount of happiness and sorrow. If one extends this individualistic conception to mankind as a whole, then it appears as if all our striving for happiness, all evolution of material conditions, all the struggle for possessions and being is a mere shifting back and forth of values whose total amount cannot be changed in this way. As a result, all changes in distribution merely reflect the basic phenomenon that one person now owns what the other, voluntarily or not, has given away. This conservation of values obviously corresponds to a pessimistic-quietistic view of life; for the less we consider ourselves able to produce really new values, the more important it is that none are really lost. The widespread notion in India that, if a holy ascetic yields to temptation his merits are transferred to the tempter, teaches this with paradoxical consistency.

But exactly opposite phenomena must also be considered. In all those emotional relationships where happiness lies not only in what one receives but just as much in what one gives, where each is mutually and equally enriched by the others, there develops a value the enjoyment of which is not bought by any deprivation on the part of an opposite party. Similarly, the communication of intellectual matters does not mean that something has to be taken from one person so that another can enjoy it. At least, only an almost pathological sensibility can bring about a feeling of deprivation if an objective intellectual idea is no longer an exclusive personal property but is also shared by others. Generally speaking, one may assert that intellectual property—at least to the extent that it does not extend into economic property—is not gained at the expense of others, since it is not taken from a limited supply but, even though its content is given, ultimately has to be produced by the thought process of whoever acquires it.

This harmonization of interests, which emanates from the nature of the object, should obviously also be provided in those economic spheres where competition for the satisfaction of individual needs is gained only at the expense of someone else. There are two types of means for transferring this situation into a more perfect one. The nearest at hand is the diversion of the struggle against fellow men towards the struggle against nature. To the extent to which further substances and forces are incorporated into human uses from the available supply of nature, competition for those that are already obtained will be reduced. . . . The extension of human spheres of power in a variety of dimensions, which belies both the statement that the world is given away free and that the satisfaction of needs is tied to theft of whatever sort, could be termed the substantive progress of culture.

Alongside this, there is what might be termed functional progress. The concern here is with finding the appropriate forms that make it advantageous for both parties to exchange ownership of specific objects. Such a form can originally have been attained only if the first owner had the physical power to keep the object wanted by others until he was offered a corresponding advantage, because otherwise the object would simply have been taken away from him. Robbery, and perhaps the gift, appear to be the most primitive stages of change in ownership, the advantage lying completely on one side and the burden falling completely on the other. When the stage of exchange appears as the form of change in ownership, or as stated earlier as a mere consequence of the equal power of both parties, then this would be evidence of the greatest progress that mankind could have made. In view of the mere differences of degree that exist in so many respects between man and the lower animals, many have often attempted to establish the specific difference that separates mankind unmistakably and unequivocally from other animals. Thus, man has been defined as the political animal, the tool-creating animal, the purposeful animal, the hierarchical animal—indeed, by a serious philosopher, as the megalomaniac animal. Perhaps we might add to this series that man is the exchanging animal, and this is in fact only one side or form of the whole general feature which seems to reflect the specific qualities of man—man is the *objective* animal. Nowhere in the animal world do we find indications of what we term objectivity, of views and treatment of things that lie beyond subjective feeling and volition.

I have already indicated how this reduces the human tragedy of competition. Such is the civilizing influence of culture that more and more contents of life become objectified in supra-individual forms: books, art, ideal concepts such as fatherland, general culture, the manifestation of life in conceptual and aesthetic images, the knowledge of a thousand interesting and significant things—all this may be enjoyed without any one depriving any other. The more values are transposed into such objective forms, the more room there is in them, as in the house of God, for every soul. Perhaps the wildness and embitterment of modern competition would be completely unbearable were it not accompanied by this growing objectivation of the contents of existence which remain untouched by all *ôte-toi que je m'y mette.*[2] It is surely of deep significance that whatever separates man on the purely factual and psychological level from lower animal species, namely the capacity for objective contemplation, the disregard of the ego with its impulses and conditions in favor of pure objectivity, contributes to the noblest and most ennobling result in the historical process: to build a world that may be acquired without conflict and mutual repression, to possess values whose acquisition and enjoyment by one person does not exclude that of another, but opens the door a thousand times for him to acquire such values as well. This problem, which is successfully solved in the world of objectivity in a substantial form, comes close to a solution in a functional form.

In contrast to the simple taking-away or gift, in which the purely subjective impulse is enjoyed, exchange presupposes, as we saw earlier, an objective appraisal, consideration, mutual acknowledgment, a restraint of direct subjective desire. It does not matter that originally this may not be voluntary but enforced by the equal power of the other party; rather, the decisive, specifically common factor is that this equivalence of power does not lead to mutual theft and struggle but to a balanced exchange in which the one-sided and personal possession or desire for

---

[2] [An attitude that says, "Get out of the way and leave me some room."]

possession enters into an objective concerted action arising out of and beyond the interaction of the subjects. Exchange—which to us appears to be something entirely self-evident—is the first, and in its simplicity really wonderful, means for combining justice with changes in ownership. Insofar as the receiver is, at the same time, the giver, the mere one-sidedness of advantage that characterizes changes of ownership dominated by a purely impulsive egotism or altruism disappears—though this does not imply that the latter relationship is always the first stage of development.

The maximization of value through changes in ownership

But the mere justice that is implied in exchange is certainly only formal and relative: any one person should have neither more nor less than any other. Over and above that, exchange brings about an increase in the absolute number of values experienced. Since everybody offers for exchange only what is relatively useless to him, and accepts in exchange what is relatively necessary, exchange effects a continuously growing utilization of the values wrested from nature at any given time. If the world were really "given away" and all activity consisted only in the mere moving back and forth of an objectively unalterable quantity of values, then exchange would nevertheless produce, as it were, an intercellular growth of values. The objectively stable sum of values changes through a more useful distribution, effected by exchange, into a subjectively larger amount and higher measure of uses experienced. This is the great cultural task of every new distribution of rights and duties, which always implies an exchange. Even in the case of an apparently quite one-sided transfer of advantages, a truly social procedure will not disregard them. Thus, for example, it was essential during the liberation of peasants in the eighteenth and nineteenth centuries not only to ensure that the landowners forfeited what the peasants were supposed to gain, but also to find a mode of distributing property and rights which enlarged the total amount of utilities.

There are two qualities of money that, in this respect, suggest that the exchange of goods or services is best served by money: its divisibility and its unlimited convertibility. The former ensures that an objective equivalence between service and its return can take place. Natural objects can seldom be so determined and scaled in value that their exchange has to be accepted as completely just by both parties. Only money—because it is nothing but the representation of the value of *other* objects, and because there is almost no limit to its divisibility and accretion—provides the technical possibility for the exact equivalence of exchange values. However, this represents only the first stage in the progressive development away from the one-sidedness of exchange of ownership.

The second quality of money derives from the fact that exchange in kind seldom gives both parties the desired object to an equal extent or is able to release them from equally superfluous ones. As a rule, the more lively desire will be on the side of one party to the transaction while the other party will enter into the exchange only by being forced to do so or where they receive a disproportionately high compensation for doing so. In the case of the exchange of services or benefits against money, however, one party receives the object that they especially need while the other receives something that anyone in general desires. Because of its boundless usefulness and therefore its permanent desirability, every exchange becomes, at least in principle, equally advantageous to both parties. The one who takes the object will certainly do so only if he needs it at this point in time; the person who takes money will accept it because he can use it at any time.

Exchange against money makes possible an increase in satisfaction for both parties, whereas with exchange in kind it is frequently the case that only one party will have a specific interest in the acquisition or disposal of the object. Thus exchange against money is so far the most perfect form of solution of the great cultural problem that evolves from the one-sided advantage of exchange of possessions, namely, to raise the objectively given amount of value to a greater amount of subjectively experienced values merely through the change in its owners. This, alongside the original creation of values, is clearly the task of social expediency as part of the general human task: to set free a maximum of the latent value that lies in the form that we give to the contents of life. Wherever we see money serving this purpose, the technical role of money also reveals that exchange is the essential social mode of solving this problem and that exchange itself is embodied in money.

The increase in the amount of satisfaction that in principle is always made possible through the commodity–money exchange process—and despite its eudaemonistic devaluation by virtue of other consequences—does not rest solely on the subjective state of one or the other parties involved in the exchange. Obviously the objective, economic fruitfulness and the intensive and extensive growth in the amount of goods in the future depends upon the manner in which any given quantity of goods is distributed at the present time. The economic consequences will be completely different depending upon who disposes of the various quantities. The mere transfer of goods from one hand to the other can subsequently considerably modify the quantity of goods in an upward or downward direction. We can even say that the same quantity of goods in different hands means a different quantity, just as the same seed in different soils produces different results. This result of the variation in distribution is most marked with regard to money. However changeable the economic importance of a landed estate or a factory may be for different owners, these variations in returns, over and above quite insignificant amounts, bear the mark of chance and abnormality. Yet the fact that the same amount of money in the hands of a stock exchange speculator or a rentier, or the State or the large industrialist produces extraordinarily different returns is a normal phenomenon that corresponds to the incomparable scope which the ownership of money provides to objective and subjective, to good and bad factors for its realization. . . .

Cultural development increases the number of persons on whom one is dependent

The importance of the money economy for individual liberty is enhanced if we explore the form that the persistent relations of dependence actually possess. As already indicated, the money economy makes possible not only a solution but a specific kind of mutual dependence which, at the same time, affords room for a maximum of liberty. Firstly, on the face of it, it creates a series of previously unknown obligations. Dependency upon third persons has spread into completely new areas ever since a considerable amount of working capital, mostly in terms of mortgages, had to be sunk into the soil in order to wrest from it the required yield. Such dependency upon third parties also spreads once tools that were directly produced with raw materials are produced indirectly by certain amounts of prefabricated components and once the laborer uses means of production which he does not own. The more the activity and life of people becomes dependent upon objective conditions by virtue of a complicated technology, the greater necessarily is the dependence upon more and more people. However, these people gain their significance for the individual concerned solely as representatives of those functions, such as owners of capital and suppliers of working materials. What kind of people they are in other respects plays no role here.

This general fact, the significance of which we shall examine later, presupposes the process by which a person acquires a definite personality in the first place. It is obviously a result of the fact that a majority of qualities, characteristic traits, and forces coalesce in a single person. Even though this person is, relatively speaking, a unity, this unity can become real and effective only by unifying a variety of determinants. Just as the essence of the physical organism lies in the fact that it creates the unity of the life-process out of the multitude of material parts, so a man's inner personal unity is based upon the interaction and connection of many elements and determinants. Each individual trait, viewed in isolation, bears an objective character; that is, it is, in and for itself, still not something personal. Neither beauty nor ugliness, neither the physical nor the intellectual centres of power, neither occupations nor intentions, nor indeed all the other innumerable human traits, unambiguously determine a personality as such. For each of them may be combined with any other quality, even with mutually incompatible elements, and may still be found in the make-up of an unlimited variety of personalities. Only the combination and fusion of several traits in one focal point forms a personality which then in its turn imports to each individual trait a personal-subjective quality. It is not that it is this *or* that trait that makes a unique personality of man, but that he is this *and* that trait. The enigmatic unity of the soul cannot be grasped by the cognitive process directly, but only when it is broken down into a multitude of strands, the re-synthesis of which signifies the unique personality.

Such a personality is almost completely destroyed under the conditions of a money economy. The delivery man, the money-lender, the worker, upon whom we are dependent, do not operate as personalities because they enter into a relationship only by virtue of a single activity such as the delivery of goods, the lending of money, and because their other qualities, which alone would give them a personality, are missing. This, of course, only signifies the ultimate stage of an on-going development which, in many ways, is not yet completed—for the dependency of human beings upon each other has not yet become wholly objectified, and personal elements have not yet been completely excluded. The general tendency, however, undoubtedly moves in the direction of making the individual more and more dependent upon the achievements of people, but less and less dependent upon the personalities that lie behind them.

Both phenomena have the same root and form the opposing sides of one and the same process: the modern division of labour permits the number of dependencies to increase just as it causes personalities to disappear behind their functions, because only one side of them operates, at the expense of all those others whose composition would make up a personality. The form of social life that would evolve were this tendency to be completely realized would exhibit a profound affinity to socialism, at least to an extreme state socialism. For socialism is concerned primarily with transforming to an extreme degree every action of social importance into an objective function. Just as today the official takes up a "position" that is objectively pre-formed and that only absorbs quite specific individual aspects or energies of his personality, so a fully fledged state socialism would erect, above the world of personalities, a world of objective forms of social action which would restrict and limit the impulses of individual personalities to very precisely and objectively determined expressions. The relationship of this world to the former is similar to that of the relationship of geometric figures to empirical bodies. The subjective tendencies and the whole of the personality could then turn into activity by restricting themselves to one-sided functional modes into which the necessary societal action is subdivided, fixed, and objectivated. The qualification of acts of the personality would thereby be completely transferred from the personality, as the *terminus a quo,*[3] to

[3] [Earliest point of origin.]

objective expediency, the *terminus ad quem*.[4] Thus, the forms of human activity would stand far above the full psychological reality of man, like the realm of Platonic ideas above the real world.

Traces of such formations do frequently exist: often a function in the division of labour confronts its holders as an independent imaginary formation so that they, no longer individually differentiated, simply pass through this function without being able or allowed to put their whole personality into these rigidly circumscribed individual demands. The personality as a mere holder of a function or position is just as irrelevant as that of a guest in a hotel room. In such a social formation, taken to its logical conclusion, the individual would be infinitely dependent; the one-sided determination of his performance would make him dependent upon supplementation by the action of all others and the satisfaction of needs would result not so much from the specific abilities of the individual but rather from an organization of work which confronted him externally and which was conceived in accordance with a completely objective standpoint. If state socialism were ever to develop to its fullest extent then it would pave the way for this differentiation of life-forms.

Money is responsible for impersonal relations between people

The money economy, however, exhibits such differentiation in the sphere of private interests. On the one hand, money makes possible the plurality of economic dependencies through its infinite flexibility and divisibility, while on the other it is conducive to the removal of the personal element from human relationships through its indifferent and objective nature. Compared with modern man, the member of a traditional or primitive economy is dependent only upon a minimum of other persons. Not only

---

[4] [Furthest possible final limit.]

is the extent of our needs considerably wider, but even the elementary necessities that we have in common with all other human beings (food, clothing, and shelter) can be satisfied only with the help of a much more complex organization and many more hands. Not only does specialization of our activities itself require an infinitely extended range of other producers with whom we exchange products, but direct action itself is dependent upon a growing amount of preparatory work, additional help, and semi-finished products. However, the relatively narrow circle of people upon whom man was dependent in an undeveloped or under-developed money economy was established much more on a personal basis. It was these specific, familiar, and at the same time irreplaceable people with whom the ancient German peasant or the Indian tribesman, the member of a Slav or Indian caste, and even medieval man frequently stood in economic relations of dependency. The fewer the number of interdependent functions, the more permanent and significant were their representatives.

In contrast, consider how many "delivery men" alone we are dependent upon in a money economy! But they are incomparably less dependent upon the specific individual and can be changed easily and frequently at any time. We have only to compare living conditions in a small town with those in a city to obtain an unmistakable though small-scale illustration of this development. While at an earlier stage man paid for the smaller number of his dependencies with the narrowness of personal relations, often with their personal irreplaceability, we are compensated for the great quantity of our dependencies by the indifference towards the respective persons and by our liberty to change them at will. And even though we are much more dependent on the whole of society through the complexity of our needs on the one hand, and the specialization of our abilities on the other, than are primitive people who could make their way through life with their very narrow isolated group, we are remarkably independent of every *specific* member of this society, because his significance for us has been transferred to the one-sided objectivity of his contribution, which can be just as easily produced by any number of other people with different personalities with whom we are connected only by an interest that can be completely expressed in money terms.

This is the most favorable situation for bringing about inner independence, the feeling of individual self-sufficiency. The mere isolation from others does not yet imply such a positive attitude. Stated in purely logical terms, independence is something other than mere non-dependence just as, say, immortality is something other than non-mortality; stone and metal are not mortal but it would not be proper to call them immortal. Even the other meaning of isolation—loneliness—reflects the erroneous impression of pure negativity. If loneliness has a psychological reality and significance then it in no way refers merely to the absence of society but rather to its ideal and then its subsequently negated existence. Loneliness is a distant effect of society, the positive determination of the individual through negative socialization. If mere isolation does not produce a longing for others or satisfaction at being remote from others—in brief, a dependency of feeling—then man is placed completely beyond the question of dependency or freedom and actual freedom takes on no conscious value because it lacks its opposite—friction, temptation, proximity to differences.

If freedom means the development of individuality, the conviction to unfold the core of our being with all its individual desires and feelings, then this category implies not a mere absence of relationships but rather a very specific relation to others. These others have to be there and to be experienced as there in order to become irrelevant. Individual freedom is not a pure inner condition of an isolated subject, but rather a phenomenon of correlation which loses its meaning when its opposite is absent. If every human relationship consists of elements of closeness and distance, then independence signifies that distance has reached a maximum, but the elements of attraction can just as little disappear altogether as can the concept of "left" exist without that of "right." The only question is then one of what is the most favorable concrete formation of both elements for promoting independence, both as an objective fact and as a subjective awareness. Such a situation seems to exist when, although extensive relations to other people exist, all genuinely individual elements have been removed from them; as in the case of mutual influences which are, however, exerted anonymously or regulations established without regard for those to whom they apply. The cause as well as the effect of such objective dependencies, where the subject as such remains free, rests upon the interchangeability of persons: the change of human subjects—voluntarily or effected by the structure of the relationship—discloses that indifference to subjective elements of dependence that characterizes the experience of freedom.

I recall the experience referred to at the beginning of this chapter, namely that a change in obligations is often experienced as freedom; it is the same form of relationship between obligations and freedom that continues here only in the individual obligation. A simple example is the characteristic difference between a medieval vassal and a serf: the vassal could change his master whereas the serf was unalterably tied to the same one. This reflects an incomparably higher measure of independence for the vassal compared with the serf, even though their obligations, considered by themselves, would have been the same. It is not the bond as such, but being bound to a particular individual master that represents the real antipode of freedom. Even the modern status of domestic servants is characterized by the fact that employers can choose by references and personal recommendations, whereas the servant has neither the chance nor the criteria for making similar choices. Only in most recent times has the scarcity of domestic servants in large cities occasionally provided the possibility of turning down a position for imponderable reasons. Both sides consider this a major step towards the independence of servants, even though the actual demands of the job are no less heavy than they previously were. If we consider the same form of relationship in an altogether different area, we can say that if an anabaptist sect justifies polygamy and the frequent change of wives on the grounds that this destroys the inner dependency or the female role then this is merely the caricature of a basically sound observation. Our overall condition is at any moment composed of both a measure of obligation and a measure of freedom, such that, within a specific sphere of life, the one is to a greater extent realized in its content, the other in its form. The restraints imposed upon us by a specific interest are felt to be less oppressive if we can choose for ourselves the objective, ideal, or personal authorities to whom we are obliged without reducing the degree of dependence.

A formally similar development emerges for wage labourers in a money economy. In view of the harshness and coerced nature of labour, it seems as if the wage labourer is nothing but a disguised slave. We shall see later how the fact that they are slaves of the objective process of production may be interpreted as a transitional stage towards their emancipation. From the subjective aspect, however, the relationship to the individual employer has become much more loose compared with earlier forms of labour. Certainly the worker is tied to his job almost as the peasant to his lot, but the frequency with which employers change in a money economy and the frequent possibility of choosing and changing them that is made possible by the form of money wages provide an altogether new freedom within the framework of his dependency. The slave could not change his master even if he had been willing to risk much worse living conditions, whereas the industrial worker can do that at any time. By thus eliminating the pressure of irrevocable dependency upon a particular individual master, the worker is already on the way to personal freedom despite his objective bondage. That this emergent freedom has little continuous influence upon the material situation of the worker should not prevent us from appreciating it. For here, as in other spheres, there is no necessary connection between liberty and increased well-being which is usually automatically presupposed by wishes, theories, and agitations.

The absence of such a connection is largely the result of the fact that the freedom of the worker is matched by the freedom of the employer which did not exist in a society of bonded labour. The slave-owner as well as the lord of the manor had a personal interest in keeping his slaves and his serfs in a good and efficient condition; his authority over them entailed his obligation to them in his own interest. Either this is not the case for the capitalist in relation to the wage labourer or, wherever it may be so, it is usually not realized. The emancipation of the labourer has to be paid for, as it were, by the emancipation of the employer, that is, by the loss of welfare that the bonded labourer enjoyed. The harshness or insecurity of his present condition is very much an indication of the process of emancipation which begins with the elimination of individually determined dependence.

Freedom in a social sense, like lack of freedom, is a relationship between human beings. Its growth implies that the relationship changes from one of stability and invariability to one of liability and interchangeability of persons. Since freedom means independence from the will of others, it commences with independence from the will of specific individuals. The lonely settler in the German or American forests is non-dependent; the inhabitants of a modern metropolis are independent in the positive sense of the word, and even though they require innumerable suppliers, workers, and co-operators, and would be lost without them, their relationship to them is completely objective and is only embodied in money. Thus the city dweller is not dependent upon any of them as particular individuals but only upon their objective services which have a money value and may therefore be carried out by any interchangeable person.

In that the purely money relationship ties the individual very closely to the group as an abstract whole and in that this is because money, in the light of our earlier deliberations, is the representative of abstract group forces, the relationship of individual persons to others simply duplicates the relationship that they have to objects as a result of money. By means of the rapid increase in the supply of commodities, on the one hand, and through the peculiar devaluation and loss of quality that objects undergo in a money economy, on the other, the individual object becomes irrelevant, often almost worthless. In contrast, not only does the whole class of these objects retain its significance, but as that culture develops our dependency on these and a steadily increasing number of objects grows; one particular pin is just as good or as worthless as any other but the modern civilized individual could not manage without pins.

Finally, the significance of money develops according to this very same tendency. The enormous cheapening of money makes specific amounts of money increasingly less valuable and important, while the role of money as a whole becomes more and more powerful and comprehensive. Within the money economy, as these phenomena illustrate, the specificity and individuality of objects becomes more and more indifferent, insubstantial, and interchangeable to us, while the actual function of the whole class of objects becomes more important and makes us increasingly dependent upon it.

This development is part of a much more general pattern which is valid for an extraordinary number of aspects and relationships of human life. These usually have their origin in an integral unity of the material and the personal. This does not mean—as it appears to us today—that the contents of life, such as property and work, duty and knowledge, social position and religion possessed some kind of independent existence, a real or conceptual independence, and that they entered any close and solidaristic union only after they had been taken up by the personality. Rather, the primary state is a complete unity, an unbroken indifference which is completely removed from the opposition of the personal and objective sides of life. . . . Thus, the contents of life—as mentioned above—develop immediately in a personal form. The emphasis on the ego on the one hand, and the object on the other, evolves from the originally naive unified form as the result of a long, never-ending process of differentiation.

The way in which the personality grows out of the state of indifference to the contents of life, the way in which, from the other side, the objectivity of things evolves, is at the same time the process of the emergence of freedom. What we term "freedom" has such a close relationship to the principle of personality that moral philosophy frequently proclaims both terms to be identical. That unity of psychic elements, their tendency to convergence in one centre, that fixed distinctness and uniqueness of the entity that we term "personality" actually means independence from an exclusion of all extraneous factors, and development exclusively according to the laws of one's own being which we call freedom. Both concepts of freedom and personality contain, in an equal measure, emphasis upon an ultimate and fundamental point in our being which stands opposed to all that is tangible, external, and sensual, both within and outside our own nature. Both are but two expressions of the single fact that a counterpart to the natural, continuous, and objectively determined existence has emerged that indicates its distinctiveness not only in the claim to an exceptional position vis-à-vis that existence but equally by striving for reconciliation with that existence. If the notion of the personality as counterpart and correlate must grow in equal measure to that of objectivity, then it becomes clear from this connection that a stricter evolution of concepts of objectivity and of individual freedom go hand in hand. Thus we can observe the distinctive parallel movement during the last three hundred years, namely that on the one hand the laws of nature, the material order of things, the objective necessity of events emerge more clearly and distinctly, while on the other we see the emphasis upon the independent individuality, upon personal freedom, upon independence in relation to all external and natural forces, becoming more and more acute and increasingly stronger. . . .

In the economy, too, the personal and objective sides of work are originally not yet separated. At first, the indifference slowly splits into opposites and the personal element increasingly recedes from production, product, and exchange. This process, however, releases individual freedom. As we have just seen, individual freedom grows to the extent that nature becomes more objective and more real for us and displays the peculiarities of its own order so that this freedom increases with the objectivation and depersonalization of the economic universe. Just as the positive sense of individual independence is not awakened in the economic isolation of an unsocial existence, then neither is it awakened in a world view which is as yet unfamiliar with the law-like regularity and strict objectivity of nature. The sense of a distinctive force and of a distinctive value of being independent is a concomitant of this opposition. Indeed, even with regard to our relationship to nature, it appears as if, in the isolation of a primitive economy—in other words, in the period of ignorance of the laws of nature in the modern sense—a much stronger bondage was reinforced by the superstitious interpretation of natural processes. Only through the growth of the economy to its full capacity, complexity, and internal interaction does the mutual dependence of people emerge. The elimination of the personal element directs the individual towards his own resources and makes him more positively aware of his liberty than would be possible with the total lack of relationships. Money is the ideal representative of such a condition since it makes possible relationships between people but leaves them personally undisturbed; it is the exact measure of material achievements, but is very inadequate for the particular and the personal. To the discriminating consciousness, the restrictedness of objective dependencies that money provides is but the background that first throws the resulting differentiated personality and its freedom into full relief.

SOPHOCLES was born in the village of Colonus near Athens, Greece in 496 B.C. His father was a prosperous munitions maker. From an early age, the talents of Sophocles were publicly recognized. At sixteen, he was selected to lead a celebration of a Greek military victory, and his civic contributions as an adult to the government of Athens were numerous. In 442, Sophocles served as a treasurer in Athens; in 440, he was elected a general; in 413, he was named an advisory commissioner. Sophocles wrote 123 tragedies, of which seven survive. Scholars believe that he also staged, choreographed, directed, and acted in these dramas when they were performed in festivals held annually in Athens. The changes made by Sophocles in the traditional form of the tragedy were lauded by Aristotle in *The Poetics.* Sophocles died in 406 B.C.

From *The Antigone of Sophocles: An English Version* by Dudley Fitts and Robert Fitzgerald. Publisher: Harcourt Brace Jovanovich, Inc., 1967.

Antigone

CHARACTERS

ANTIGONE, ISMENE — daughters of Oedipus

CREON, King of Thebes

EURYDICE, his wife

HAIMON, his son

TEIRESIAS, the blind prophet

A SENTRY

A MESSENGER

CHORUS, citizens of Thebes

CHORAGOS, leader of the CHORUS

SCENE: *Before the Palace of* CREON, *King of Thebes. A central double door, and two lateral doors. A platform extends the length of the facade, and from this platform three steps lead down into the orchestra, or chorus-ground. Time: dawn of the day after the repulse of the Argive army from the assault on Thebes.*

PROLOGUE

(ANTIGONE *and* ISMENE *enter from the central door of the Palace.*)

ANTIGONE: Ismene, dear sister,
You would think that we had already suffered enough
For the curse on Oedipus:
I cannot imagine any grief
That you and I have not gone through. And now—
Have they told you the new decree of our King Creon?

ISMENE: I have heard nothing: I know
That two sisters lost two brothers, a double death
In a single hour; and I know that the Argive army
Fled in the night; but beyond this, nothing.

ANTIGONE: I thought so. And that is why I wanted you
To come out here with me. There is something we must do.

ISMENE: Why do you speak so strangely?

ANTIGONE: Listen, Ismene:
Creon buried our brother Eteocles
With military honors, gave him a soldier's funeral,
And it was right that he should; but Polyneices,
Who fought as bravely and died as miserably,—
They say that Creon has sworn
No one shall bury him, no one mourn for him,
But his body must lie in the fields, a sweet treasure
For carrion birds to find as they search for food.
That is what they say, and our good Creon is coming here
To announce it publicly; and the penalty—

Stoning to death in the public square!
There it is,
And now you can prove what you are:
A true sister, or a traitor to your family.

ISMENE: Antigone, you are mad! What could I possibly do?

ANTIGONE: You must decide whether you will help me or not.

ISMENE: I do not understand you. Help you in what?

ANTIGONE: Ismene, I am going to bury him. Will you come?

ISMENE: Bury him! You have just said the new law forbids it.

ANTIGONE: He is my brother. And he is your brother, too.

ISMENE: But think of the danger! Think what Creon will do!

ANTIGONE: Creon is not strong enough to stand in my way.

ISMENE: Ah sister!
Oedipus died, everyone hating him
For what his own search brought to light, his eyes
Ripped out by his own hand; and Iocaste died,
His mother and wife at once: she twisted the cords
That strangled her life; and our two brothers died,
Each killed by the other's sword. And we are left:
But oh, Antigone,
Think how much more terrible than these
Our own death would be if we should go against Creon
And do what he has forbidden! We are only women,
We cannot fight with men, Antigone!
The law is strong, we must give in to the law
In this thing, and in worse. I beg the Dead
To forgive me, but I am helpless: I must yield
To those in authority. And I think it is dangerous business
To be always meddling.

ANTIGONE: If that is what you think,
I should not want you, even if you asked to come.

You have made your choice, you can be what you want to be.
But I will bury him; and if I must die,
I say that this crime is holy: I shall lie down
With him in death, and I shall be as dear
To him as he to me.
It is the dead,
Not the living, who make the longest demands:
We die forever . . .
You may do as you like,
Since apparently the laws of the gods mean nothing to you.

ISMENE: They mean a great deal to me; but I have no strength
To break laws that were made for the public good.

ANTIGONE: That must be your excuse, I suppose. But as for me,
I will bury the brother I love.

ISMENE: Antigone,
I am so afraid for you!

ANTIGONE: You need not be:
You have yourself to consider, after all.

ISMENE: But no one must hear of this, you must tell no one!
I will keep it a secret, I promise!

ANTIGONE: Oh tell it! Tell everyone!
Think how they'll hate you when it all comes out
If they learn that you knew about it all the time!

ISMENE: So fiery! You should be cold with fear.

ANTIGONE: Perhaps. But I am doing only what I must.

ISMENE: But can you do it? I say that you cannot.

ANTIGONE: Very well: when my strength gives out, I shall do no more.

ISMENE: Impossible things should not be tried at all.

ANTIGONE: Go away, Ismene:
I shall be hating you soon, and the dead will too,
For your words are hateful. Leave me my foolish plan:
I am not afraid of the danger; if it means death,
It will not be the worst of deaths—death without honor.

#13

ISMENE: Go then, if you feel that you must.
You are unwise,
But a loyal friend indeed to those who love you.

(*Exit into the Palace.* ANTIGONE *goes off.*
Enter the CHORUS.)

PARODOS

CHORUS: Now the long blade of the sun, lying
Level east to west, touches with glory
Thebes of the Seven Gates. Open, unlidded
Eye of golden day! O marching light
Across the eddy and rush of Dirce's stream,
Striking the white shields of the enemy
Thrown headlong backward from the blaze of morning!

CHORAGOS: Polyneices their commander
Roused them with windy phrases,
He the wild eagle screaming
Insults above our land,
His wings their shields of snow,
His crest their marshalled helms.

CHORUS: Against our seven gates in a yawning ring
The famished spears came onward in the night;
But before his jaws were sated with our blood,
Or pinefire took the garland of our towers,
He was thrown back; and as he turned, great Thebes—
No tender victim for his noisy power—
Rose like a dragon behind him, shouting war.

CHORAGOS: For God hates utterly
The bray of bragging tongues;
And when he beheld their smiling,
Their swagger of golden helms,
The frown of his thunder blasted
Their first man from our walls.

CHORUS: We heard his shout of triumph high in the air
Turn to a scream; far out in a flaming arc
He fell with his windy torch, and the earth struck him.
And others storming in fury no less than his
Found shock of death in the dusty joy of battle.

CHORAGOS: Seven captains at seven gates
Yielded their clanging arms to the god
That bends the battle-line and breaks it.
These two only, brothers in blood,
Face to face in matchless rage,
Mirroring each the other's death,
Clashed in long combat.

CHORUS: But now in the beautiful morning of victory
Let Thebes of the many chariots sing for joy!
With hearts for dancing we'll take leave of war:
Our temples shall be sweet with hymns of praise,
And the long night shall echo with our chorus.

SCENE ONE

CHORAGOS: But now at last our new King is coming:
Creon of Thebes, Menoiceus' son.
In this auspicious dawn of his reign
What are the new complexities
That shifting Fate has woven for him?
What is his counsel? Why has he summoned
The old men to hear him?

(*Enter* CREON *from the Palace.*
He addresses the CHORUS *from the top step.*)

CREON: Gentlemen: I have the honor to inform you that our Ship of State, which recent storms have threatened to destroy, has come safely to harbor at last, guided by the merciful wisdom of Heaven. I have summoned you here this morning because I know that I can depend upon you: your devotion to King Laios was absolute; you never hesitated in your duty to our late ruler Oedipus; and when Oedipus died, your loyalty was transferred to his children. Unfortunately, as you know, his two sons, the princes Eteocles and Polyneices, have killed each other in battle; and I, as the next in blood, have succeeded to the full power of the throne.

I am aware, of course, that no Ruler can expect complete loyalty from his subjects until he has been tested in office. Nevertheless, I say to you at the very outset that I have nothing but contempt for the kind of Governor who is afraid, for whatever reason, to follow the course that he knows is best for the State; and as for the man who sets private friendship above the public welfare,—I have no use for him, either. I call God to witness that if I saw my country headed for ruin, I should not be afraid to speak out plainly; and I need hardly remind you that I would never have any dealings with an enemy of the people. No one values friendship more highly than I; but we must remember that friends made at the risk of wrecking our Ship are not real friends at all.

These are my principles, at any rate, and that is why I have made the following decision concerning the sons of Oedipus: Eteocles, who died as a man should die, fighting for his country, is to be buried with full military honors, with all the ceremony that is usual when the greatest heroes die; but his brother Polyneices, who broke his exile to come back with fire and sword against his native city and the shrines of his fathers' gods, whose own idea was to spill the blood of his blood and sell his own people into slavery—Polyneices, I say, is to have no burial: no man is to touch him or say the least prayer for him; he shall lie on the plain, unburied; and the birds and the scavenging dogs can do with him whatever they like.

This is my command, and you can see the wisdom behind it. As long as I am King, no traitor is going to be honored with the loyal man. But whoever shows by word and deed that he is on the side of the State—he shall have my respect while he is living, and my reverence when he is dead.

CHORAGOS: If that is your will, Creon son of Menoiceus,
You have the right to enforce it: we are yours.

CREON: That is my will. Take care that you do your part.

CHORAGOS: We are old men: let the younger ones carry it out.

CREON: I do not mean that: the sentries have been appointed.

CHORAGOS: Then what is it that you would have us do?

CREON: You will give no support to whoever breaks this law.

CHORAGOS: Only a crazy man is in love with death!

CREON: And death it is: yet money talks, and the wisest
Have sometimes been known to count a few coins too many.

(*Enter* SENTRY.)

SENTRY: I'll not say that I'm out of breath from running, King, because every time I stopped to think about what I have to tell you, I felt like going back. And all the time a voice kept saying, "You fool, don't you know you're walking straight into trouble?"; and then another voice: "Yes, but if you let somebody else get the news to Creon first, it will be even worse than that for you!" But good sense won out, at least I hope it was good sense, and here I am with a story that makes no sense at all; but I'll tell it anyhow, because, as they say, what's going to happen's going to happen, and—

CREON: Come to the point. What have you to say?

SENTRY: I did not do it. I did not see who did it. You must not punish me for what someone else has done.

CREON: A comprehensive defense! More effective, perhaps,
If I knew its purpose. Come: what is it?

SENTRY: A dreadful thing . . . I don't know how to put it—

CREON: Out with it!

SENTRY: Well, then;
The dead man—
Polyneices—
out there—
someone—
New dust on the slimy flesh!
Someone has given it burial that way, and
Gone . . .

CREON: And the man who dared do this?

SENTRY: I swear I
Do not know! You must believe me!
Listen:
The ground was dry, not a sign of digging, no,
Not a wheeltrack in the dust, no trace of anyone.
It was when they relieved us this morning: and one of them,
The corporal, pointed to it.
There it was,
The strangest—
Look:
The body, just mounded over there with light dust: you see?
Not buried really, but as if they'd covered it
Just enough for the ghost's peace. And no sign
Of dogs or any wild animal that had been there.

And then what a scene there was! Every man of us
Accusing the other: we all proved the other man did it,
We all had proof that we could not have done it.
We were ready to take hot iron in our hands,
Walk through fire, swear by all the gods,
It was not I!
I do not know who it was, but it was not I!
And then, when this came to nothing, someone said
A thing that silenced us and made us stare
Down at the ground: you had to be told the news,
And one of us had to do it! We threw the dice,
And the bad luck fell to me. So here I am,
No happier to be here than you are to have me:
Nobody likes the man who brings bad news.

CHORAGOS: I have been wondering, King: can it be that the gods have done this?

CREON: Stop!
Must you doddering wrecks
Go out of your heads entirely? "The gods!"
Intolerable!
The gods favor this corpse? Why? How had he served them?
Tried to loot their temples, burn their images,
Yes, and the whole State, and its laws with it!
Is it your senile opinion that the gods love to honor bad men?
A pious thought!—
No, from the very beginning
There have been those who have whispered together,
Stiff-necked anarchists, putting their heads together,
Scheming against me in alleys. These are the men,
And they have bribed my own guard to do this thing.

Money!
There's nothing in the world so demoralizing as money.
Down go your cities,
Homes gone, men gone, honest hearts corrupted,

Crookedness of all kinds, and all for money!
(*To* SENTRY.) But you—!
I swear by God and by the throne of God,
The man who has done this thing shall pay for it!
Find that man, bring him here to me, or your death
Will be the least of your problems: I'll string you up
Alive, and there will be certain ways to make you
Discover your employer before you die;
And the process may teach you a lesson you seem to have missed:
The dearest profit is sometimes all too dear:
That depends on the source. Do you understand me?
A fortune won is often misfortune.

SENTRY: King, may I speak?

CREON: Your very voice distresses me.

SENTRY: Are you sure that it is my voice, and not your conscience?

CREON: By God, he wants to analyze me now!

SENTRY: It is not what I say, but what has been done, that hurts you.

CREON: You talk too much.

SENTRY: Maybe; but I've done nothing.

CREON: Sold your soul for some silver; that's all you've done.

SENTRY: How dreadful it is when the right judge judges wrong!

CREON: Your figures of speech
May entertain you now; but unless you bring me the man,
You will get little profit from them in the end.

(*Exit* CREON *into the Palace.*)

SENTRY: "Bring me the man"—!
I'd like nothing better than bringing him the man!
But bring him or not, you have seen the last of me here.
At any rate, I am safe! (*Exit* SENTRY.)

ODE I

CHORUS: Numberless are the world's wonders, but none
More wonderful than man; the stormgray sea
Yields to his prows, the huge crests bear him high;
Earth, holy and inexhaustible, is graven
With shining furrows where his plows have gone
Year after year, the timeless labor of stallions.

The lightboned birds and beasts that cling to cover,
The lithe fish lighting their reaches of dim water,
All are taken, tamed in the net of his mind;
The lion on the hill, the wild horse windy-maned,
Resign to him; and his blunt yoke has broken
The sultry shoulders of the mountain bull.

Words also, and thought as rapid as air,
He fashions to his good use; statecraft is his,
And his the skill that deflects the arrows of snow,
The spears of winter rain: from every wind
He has made himself secure—from all but one:
In the late wind of death he cannot stand.

O clear intelligence, force beyond all measure!
O fate of man, working both good and evil!
When the laws are kept, how proudly his city stands!
When the laws are broken, what of his city then?
Never may the anarchic man find rest at my hearth,
Never be it said that my thoughts are his thoughts.

SCENE TWO

(*Re-enter* SENTRY *leading* ANTIGONE.)

CHORAGOS: What does this mean? Surely this captive woman
Is the Princess, Antigone. Why should she be taken?

SENTRY: Here is the one who did it! We caught her
In the very act of burying him.—Where is Creon?

CHORAGOS: Just coming from the house. (*Enter* CREON.)

CREON: What has happened?
Why have you come back so soon?

SENTRY: O King,
A man should never be too sure of anything:
I would have sworn
That you'd not see me here again: your anger
Frightened me so, and the things you threatened me with;
But how could I tell then
That I'd be able to solve the case so soon?
No dice-throwing this time: I was only too glad to come!
Here is this woman. She is the guilty one:
We found her trying to bury him.
Take her, then; question her; judge her as you will.
I am through with the whole thing now, and glad of it.

CREON: But this is Antigone! Why have you brought her here?

SENTRY: She was burying him, I tell you!

CREON: Is this the truth?

SENTRY: I saw her with my own eyes. Can I say more?

CREON: The details: come, tell me quickly!

SENTRY: It was like this:
After those terrible threats of yours, King,
We went back and brushed the dust away from the body.
The flesh was soft by now, and stinking,
So we sat on a hill to windward and kept guard.
No napping this time! We kept each other awake.
But nothing happened until the white round sun
Whirled in the center of the round sky over us:
Then, suddenly,
A storm of dust roared up from the earth, and the sky
Went out, the plain vanished with all its trees
In the stinging dark. We closed our eyes and endured it.

The whirlwind lasted a long time, but it passed;
And then we looked, and there was Antigone!
I have seen
A mother bird come back to a stripped nest, heard
Her crying bitterly a broken note or two
For the young ones stolen. Just so, when this girl
Found the bare corpse, and all her love's work wasted,
She wept, and cried on heaven to damn the hands
That had done this thing.
And then she brought more dust
And sprinkled wine three times for her brother's ghost.
We ran and took her at once. She was not afraid,
Not even when we charged her with what she had done.
She denied nothing.
And this was a comfort to me,
And some uneasiness: for it is a good thing
To escape from death, but it is no great pleasure
To bring death to a friend.
Yet I always say
There is nothing so comfortable as your own safe skin!

CREON: And you, Antigone,
You with your head hanging—do you confess this thing?

ANTIGONE: I do. I deny nothing.

CREON (*to* SENTRY): You may go. (*Exit* SENTRY.)
(*To* ANTIGONE.) Tell me, tell me briefly:
Had you heard my proclamation touching this matter?

ANTIGONE: It was public. Could I help hearing it?

CREON: And yet you dared defy the law.

ANTIGONE: I dared.
It was not God's proclamation. That final Justice
That rules the world below makes no such laws.
Your edict, King, was strong,
But all your strength is weakness itself against

The immortal unrecorded laws of God.
They are not merely now: they were, and shall be,
Operative forever, beyond man utterly.
I knew I must die, even without your decree:
I am only mortal. And if I must die
Now, before it is my time to die,
Surely this is no hardship: can anyone
Living, as I live, with evil all about me,
Think Death less than a friend? This death of mine
Is of no importance; but if I had left my brother
Lying in death unburied, I should have suffered.
Now I do not.
You smile at me. Ah Creon,
Think me a fool, if you like; but it may well be
That a fool convicts me of folly.

CHORAGOS: Like father, like daughter: both headstrong, deaf to reason!
She has never learned to yield.

CREON: She has much to learn.
The inflexible heart breaks first, the toughest iron
Cracks first, and the wildest horses bend their necks
At the pull of the smallest curb.
Pride? In a slave?
This girl is guilty of a double insolence,
Breaking the given laws and boasting of it.
Who is the man here,
She or I, if this crime goes unpunished?
Sister's child, or more than sister's child,
Or closer yet in blood—she and her sister
Win bitter death for this!
(*To* Servants.)
Go, some of you,
Arrest Ismene. I accuse her equally.
Bring her: you will find her sniffling in the house there.
Her mind's a traitor: crimes kept in the dark

Cry for light, and the guardian brain shudders;
But how much worse than this
Is brazen boasting of barefaced anarchy!

ANTIGONE: Creon, what more do you want than my death?

CREON: Nothing.
That gives me everything.

ANTIGONE: Then I beg you: kill me.
This talking is a great weariness: your words
Are distasteful to me, and I am sure that mine
Seem so to you. And yet they should not seem so:
I should have praise and honor for what I have done.
All these men here would praise me
Were their lips not frozen shut with fear of you.
Ah the good fortune of kings,
Licensed to say and do whatever they please!

CREON: You are alone here in that opinion.

ANTIGONE: No, they are with me. But they keep their tongues in leash.

CREON: Maybe. But you are guilty, and they are not.

ANTIGONE: There is no guilt in reverence for the dead.

CREON: But Eteocles—was he not your brother too?

ANTIGONE: My brother too.

CREON: And you insult his memory?

ANTIGONE: The dead man would not say that I insult it.

CREON: He would: for you honor a traitor as much as him.

ANTIGONE: His own brother, traitor or not, and equal in blood.

CREON: He made war on his country. Eteocles defended it.

ANTIGONE: Nevertheless, there are honors due all the dead.

CREON: But not the same for the wicked as for the just.

ANTIGONE: Ah Creon, Creon,
Which of us can say what the gods hold wicked?

CREON: An enemy is an enemy, even dead.

ANTIGONE: It is my nature to join in love, not hate.

CREON: Go join them, then; if you must have your love,
Find it in hell!

CHORAGOS: But see, Ismene comes:

(*Enter* ISMENE, *guarded.*)

Those tears are sisterly, the cloud
That shadows her eyes rains down gentle sorrow.

CREON: You too, Ismene,
Snake in my ordered house, sucking my blood
Stealthily—and all the time I never knew
That these two sisters were aiming at my throne!
Ismene,
Do you confess your share in this crime, or deny it?
Answer me.

ISMENE: Yes, if she will let me say so. I am guilty.

ANTIGONE: No, Ismene. You have no right to say so.
You would not help me, and I will not have you help me.

ISMENE: But now I know what you meant; and I am here
To join you, to take my share of punishment.

ANTIGONE: The dead man and the gods who rule the dead
Know whose act this was. Words are not friends.

ISMENE: Do you refuse me, Antigone? I want to die with you:
I too have a duty that I must discharge to the dead.

ANTIGONE: You shall not lessen my death by sharing it.

ISMENE: What do I care for life when you are dead?

ANTIGONE: Ask Creon. You're always hanging on his opinions.

ISMENE: You are laughing at me. Why, Antigone?

ANTIGONE: It's a joyless laughter, Ismene.

ISMENE: But can I do nothing?

ANTIGONE: Yes. Save yourself. I shall not envy you.
There are those who will praise you; I shall have honor, too.

ISMENE: But we are equally guilty!

ANTIGONE: No more, Ismene.
You are alive, but I belong to Death.

CREON (*to the* CHORUS): Gentlemen, I beg you to observe these girls:
One has just now lost her mind; the other,
It seems, has never had a mind at all.

ISMENE: Grief teaches the steadiest minds to waver, King.

CREON: Yours certainly did, when you assumed guilt with the guilty!

ISMENE: But how could I go on living without her?

CREON: You are.
She is already dead.

ISMENE: But your own son's bride!

CREON: There are places enough for him to push his plow.
I want no wicked women for my sons!

ISMENE: O dearest Haimon, how your father wrongs you!

CREON: I've had enough of your childish talk of marriage!

CHORAGOS: Do you really intend to steal this girl from your son?

CREON: No; Death will do that for me.

CHORAGOS: Then she must die?

CREON: You dazzle me.

—But enough of this talk!

(*To* Guards.)

You, there, take them away and guard them well:
For they are but women, and even brave men run
When they see Death coming.

(*Exeunt* ISMENE, ANTIGONE, *and* Guards.)

ODE II

CHORUS: Fortunate is the man who has never tasted God's vengeance!
Where once the anger of heaven has struck, that house is shaken
Forever: damnation rises behind each child
Like a wave cresting out of the black northeast,
When the long darkness under sea roars up
And bursts drumming death upon the wind-whipped sand.

I have seen this gathering sorrow from time long past
Loom upon Oedipus' children: generation from generation
Takes the compulsive rate of the enemy god.
So lately this last flower of Oedipus' line
Drank the sunlight! but now a passionate word
And a handful of dust have closed up all its beauty.

What mortal arrogance
Transcends the wrath of Zeus?
Sleep cannot lull him, nor the effortless long months
Of the timeless gods: but he is young forever,

And his house is the shining day of high Olympos.
All that is and shall be,
And all the past, is his.
No pride on earth is free of the curse of heaven.

The straying dreams of men
May bring them ghosts of joy:
But as they drowse, the waking embers burn them;
Or they walk with fixed eyes, as blind men walk.
But the ancient wisdom speaks for our own time:
Fate works most for woe
With Folly's fairest show.
Man's little pleasure is the spring of sorrow.

SCENE THREE

CHORAGOS: But here is Haimon, King, the last of all your sons.
Is it grief for Antigone that brings him here,
And bitterness at being robbed of his bride?

(*Enter* HAIMON.)

CREON: We shall soon see, and no need of diviners.
—Son,
You have heard my final judgment on that girl:
Have you come here hating me, or have you come
With deference and with love, whatever I do?

HAIMON: I am your son, father. You are my guide.
You make things clear for me, and I obey you.
No marriage means more to me than your continuing wisdom.

CREON: Good. That is the way to behave: subordinate
Everything else, my son, to your father's will.
This is what a man prays for, that he may get
Sons attentive and dutiful in his house,
Each one hating his father's enemies,
Honoring his father's friends. But if his sons

Fail him, if they turn out unprofitably,
What has he fathered but trouble for himself
And amusement for the malicious?
                                                            So you are right
Not to lose your head over this woman.
Your pleasure with her would soon grow cold, Haimon,
And then you'd have a hellcat in bed and elsewhere.
Let her find her husband in Hell!
Of all the people in this city, only she
Has had contempt for my law and broken it.

Do you want me to show myself weak before the people?
Or to break my sworn word? No, and I will not.
The woman dies.

I suppose she'll plead "family ties." Well, let her.
If I permit my own family to rebel,
How shall I earn the world's obedience?
Show me the man who keeps his house in hand,
He's fit for public authority.
                                                    I'll have no dealings
With law-breakers, critics of the government:
Whoever is chosen to govern should be obeyed—
Must be obeyed, in all things, great and small,
Just and unjust! O Haimon,
The man who knows how to obey, and that man only,
Knows how to give commands when the time comes.
You can depend on him, no matter how fast
The spears come: he's a good soldier, he'll stick it out.

Anarchy, anarchy! Show me a greater evil!
This is why cities tumble and the great houses rain down,
This is what scatters armies!

No, no: good lives are made so by discipline.
We keep the laws then, and the lawmakers,

And no woman shall seduce us. If we must lose,
Let's lose to a man, at least! Is a woman stronger than we?

CHORAGOS: Unless time has rusted my wits,
What you say, King, is said with point and dignity.

HAIMON: Father:
Reason is God's crowning gift to man, and you are right
To warn me against losing mine. I cannot say—
I hope that I shall never want to say!—that you
Have reasoned badly. Yet there are other men
Who can reason, too; and their opinions might be helpful.
You are not in a position to know everything
That people say or do, or what they feel:
Your temper terrifies them—everyone
Will tell you only what you like to hear.
But I, at any rate, can listen; and I have heard them
Muttering and whispering in the dark about this girl.
They say no woman has ever, so unreasonably,
Died so shameful a death for a generous act:
"She covered her brother's body. Is this indecent?
She kept him from dogs and vultures. Is this a crime?
Death?—She should have all the honor that we can give her!"

This is the way they talk out there in the city.

You must believe me:
Nothing is closer to me than your happiness.
What could be closer? Must not any son
Value his father's fortune as his father does his?
I beg you, do not be unchangeable:
Do not believe that you alone can be right.
The man who thinks that,
The man who maintains that only he has the power
To reason correctly, the gift to speak, the soul—
A man like that, when you know him, turns out empty.

It is not reason never to yield to reason!

In flood time you can see how some trees bend,
And because they bend, even their twigs are safe,
While stubborn trees are torn up, roots and all.
And the same thing happens in sailing:
Make your sheet fast, never slacken,—and over you go.
Head over heels and under: and there's your voyage.
Forget you are angry! Let yourself be moved!
I know I am young; but please let me say this:
The ideal condition
Would be, I admit, that men should be right by instinct;
But since we are all too likely to go astray,
The reasonable thing is to learn from those who can teach.

#9
CHORAGOS: You will do well to listen to him, King,
If what he says is sensible. And you, Haimon,
Must listen to your father.—Both speak well.

CREON: You consider it right for a man of my years and experience
To go to school to a boy?

HAIMON: It is not right
If I am wrong. But if I am young, and right,
What does my age matter?

CREON: You think it right to stand up for an anarchist?

HAIMON: Not at all. I pay no respect to criminals.

CREON: Then she is not a criminal?

HAIMON: The City would deny it, to a man.

CREON: And the City proposes to teach me how to rule?

HAIMON: Ah. Who is it that's talking like a boy now?

CREON: My voice is the one voice giving orders in this City!

HAIMON: It is no City if it takes orders from one voice.

CREON: The State is the King!

HAIMON: Yes, if the State is a desert.

CREON: This boy, it seems, has sold out to a woman.

HAIMON: If you are a woman: my concern is only for you.

CREON: So? Your "concern"! In a public brawl with your father!

HAIMON: How about you, in a public brawl with justice?

CREON: With justice, when all that I do is within my rights?

HAIMON: You have no right to trample on God's right.

CREON: Fool, adolescent fool! Taken in by a woman!

HAIMON: You'll never see me taken in by anything vile.

CREON: Every word you say is for her!

HAIMON: And for you.
And for me. And for the gods under the earth.

CREON: You'll never marry her while she lives.

HAIMON: Then she must die.—But her death will cause another.

CREON: Another?
Have you lost your senses? Is this an open threat?

HAIMON: There is no threat in speaking to emptiness.

CREON: I swear you'll regret this superior tone of yours!
You are the empty one!

HAIMON: If you were not my father,
I'd say you were perverse.

CREON: You girlstruck fool, don't play at words with me!

HAIMON: I am sorry. You prefer silence.

CREON: Now, by God—!
I swear, by all the gods in heaven above us,
You'll watch it, I swear you shall!
(*To the* SERVANTS.) Bring her out!
Bring the woman out! Let her die before his eyes!
Here, this instant, with her bridegroom beside her!

HAIMON: Not here, no; she will not die here, King.
And you will never see my face again.
Go on raving as long as you've a friend to endure you.

(*Exit* HAIMON.)

CHORAGOS: Gone, gone.
Creon, a young man in a rage is dangerous!

CREON: Let him do, or dream to do, more than a man can.
He shall not save these girls from death.

CHORAGOS: These girls?
You have sentenced them both?

CREON: No, you are right.
I will not kill the one whose hands are clean.

CHORAGOS: But Antigone?

CREON: I will carry her far away
Out there in the wilderness, and lock her
Living in a vault of stone. She shall have food,
As the custom is, to absolve the State of her death.
And there let her pray to the gods of hell:
They are her only gods:
Perhaps they will show her an escape from death,
Or she may learn,
though late,
That piety shown the dead is piety in vain.

(*Exit* CREON.)

ODE III

CHORUS: Love, unconquerable
Waster of rich men, keeper
Of warm lights and all-night vigil
In the soft face of a girl:
Sea-wanderer, forest-visitor!
Even the pure Immortals cannot escape you,
And mortal man, in his one day's dusk,
Trembles before your glory.

Surely you swerve upon ruin
The just man's consenting heart,
As here you have made bright anger
Strike between father and son—
And none has conquered but Love!
A girl's glance working the will of heaven:
Pleasure to her alone who mocks us,
Merciless Aphrodite.

SCENE FOUR

(ANTIGONE *enters, guarded.*)

CHORAGOS: But I can no longer stand in awe of this,
Nor, seeing what I see, keep back my tears.
Here is Antigone, passing to that chamber
Where all find sleep at last.

ANTIGONE: Look upon me, friends, and pity me
Turning back at the night's edge to say
Good-bye to the sun that shines for me no longer;
Now sleepy Death
Summons me down to Acheron, that cold shore:
There is no bridesong there, nor any music.

CHORUS: Yet not unpraised, not without a kind of honor,
You walk at last into the underworld;
Untouched by sickness, broken by no sword.
What woman has ever found your way to death?

ANTIGONE: How often I have heard the story of Niobe,
Tantalos' wretched daughter, how the stone
Clung fast about her, ivy-close: and they say
The rain falls endlessly
And sifting soft snow; her tears are never done.
I feel the loneliness of her death in mine.

CHORUS: But she was born of heaven, and you
Are woman, woman-born. If her death is yours,
A mortal woman's, is this not for you
Glory in our world and in the world beyond?

ANTIGONE: You laugh at me. Ah, friends, friends,
Can you not wait until I am dead? O Thebes,
O men many-charioted, in love with Fortune,
Dear springs of Dirce, sacred Theban grove,
Be witnesses for me, denied all pity,
Unjustly judged! and think a word of love
For her whose path turns
Under dark earth, where there are no more tears.

CHORUS: You have passed beyond human daring and come at last
Into a place of stone where Justice sits.
I cannot tell
What shape of your father's guilt appears in this.

ANTIGONE: You have touched it at last: that bridal bed
Unspeakable, horror of son and mother mingling:
#1 Their crime, infection of all our family!
O Oedipus, father and brother!
Your marriage strikes from the grave to murder mine.
I have been a stranger here in my own land:

All my life
The blasphemy of my birth has followed me.

CHORUS: Reverence is a virtue, but strength
Lives in established law: that must prevail.
You have made your choice,
Your death is the doing of your conscious hand.

ANTIGONE: Then let me go, since all your words are bitter,
And the very light of the sun is cold to me.
Lead me to my vigil, where I must have
Neither love nor lamentation; no song, but silence.

CREON: If dirges and planned lamentations could put off death,
Men would be singing forever.
(*To the* Servants.)
Take her, go!
You know your orders: take her to the vault
And leave her alone there. And if she lives or dies,
That's her affair, not ours: our hands are clean.

ANTIGONE: O tomb, vaulted bride-bed in eternal rock,
Soon I shall be with my own again
Where Persephone welcomes the thin ghosts underground:
And I shall see my father again, and you, mother,
And dearest Polyneices—
dearest indeed
To me, since it was my hand
That washed him clean and poured the ritual wine:
And my reward is death before my time!

And yet, as men's hearts know, I have done no wrong,
I have not sinned before God. Or if I have,
I shall know the truth in death. But if the guilt
Lies upon Creon who judged me, then, I pray,
May his punishment equal my own.

CHORAGOS: O passionate heart,
Unyielding, tormented still by the same winds!

CREON: Her guards shall have good cause to regret their delaying.

ANTIGONE: Ah! That voice is like the voice of death!

CREON: I can give you no reason to think you are mistaken.

ANTIGONE: Thebes, and you my fathers' gods,
And rulers of Thebes, you see me now, the last
Unhappy daughter of a line of kings,
Your kings, led away to death. You will remember
What things I suffer, and at what men's hands,
Because I would not transgress the laws of heaven.
(*To the* Guards.)
Come: let us wait no longer.

(*Exit* ANTIGONE, *guarded.*)

ODE IV

CHORUS: All Danae's beauty was locked away
In a brazen cell where the sunlight could not come:
A small room, still as any grave, enclosed her.
Yet she was a princess too,
And Zeus in a rain of gold poured love upon her.
O child, child,
No power in wealth or war
Or tough sea-blackened ships
Can prevail against untiring Destiny!

And Dryas' son also, that furious king,
Bore the god's prisoning anger for his pride:
Sealed up by Dionysos in deaf stone,
His madness died among echoes.
So at the last he learned what dreadful power
His tongue had mocked:

For he had profaned the revels,
And fired the wrath of the nine
Implacable Sisters that love the sound of the flute.

And old men tell a half-remembered tale
Of horror done where a dark ledge splits the sea
And a double surf beats on the gray shores:
How a king's new woman, sick
With hatred for the queen he had imprisoned,
Ripped out his two sons' eyes with her bloody hands
While grinning Ares watched the shuttle plunge
Four times: four blind wounds crying for revenge,

Crying, tears and blood mingled.—Piteously born,
Those sons whose mother was of heavenly birth!
Her father was the god of the North Wind
And she was cradled by gales,
She raced with young colts on the glittering hills
And walked untrammeled in the open light:
But in her marriage deathless Fate found means
To build a tomb like yours for all her joy.

SCENE FIVE

(*Enter blind* TEIRESIAS, *led by a boy.*)

TEIRESIAS: This is the way the blind man comes, Princes, Princes,
Lock-step, two heads lit by the eyes of one.

CREON: What new thing have you to tell us, old Teiresias?

TEIRESIAS: I have much to tell you: listen to the prophet, Creon.

CREON: I am not aware that I have ever failed to listen.

TEIRESIAS: Then you have done wisely, King, and ruled well.

CREON: I admit my debt to you. But what have you to say?

TEIRESIAS: This, Creon: you stand once more on the edge of fate.

CREON: What do you mean? Your words are a kind of dread.

TEIRESIAS: Listen, Creon:
I was sitting in my chair of augury, at the place
Where the birds gather about me. They were all a-chatter,
As is their habit, when suddenly I heard
A strange note in their jangling, a scream, a
Whirring fury; I knew that they were fighting,
Tearing each other, dying
In a whirlwind of wings clashing. And I was afraid.
I began the rites of burnt-offering at the altar,
But Hephaistos failed me: instead of bright flame,
There was only the sputtering slime of the fat thigh-flesh
Melting: the entrails dissolved in gray smoke,
The bare bone burst from the welter. And no blaze!

This was a sign from heaven. My boy described it,
Seeing for me as I see for others.

I tell you, Creon, you yourself have brought
This new calamity upon us. Our hearths and altars
Are stained with the corruption of dogs and carrion birds
That glut themselves on the corpse of Oedipus' son.
The gods are deaf when we pray to them, their fire
Recoils from our offering, their birds of omen
Have no cry of comfort, for they are gorged
With the thick blood of the dead.
O my son,
These are no trifles! Think: all men make mistakes,
But a good man yields when he knows his course is wrong,
And repairs the evil. The only crime is pride.

Give in to the dead man, then: do not fight with a corpse—
What glory is it to kill a man who is dead?
Think, I beg you:

It is for your own good that I speak as I do.
You should be able to yield for your own good.

CREON: It seems that prophets have made me their especial province.
All my life long
I have been a kind of butt for the dull arrows
Of doddering fortune-tellers!
No, Teiresias:
If your birds—if the great eagles of God himself
Should carry him stinking bit by bit to heaven,
I would not yield. I am not afraid of pollution:
No man can defile the gods.
Do what you will,
Go into business, make money, speculate
In India gold or that synthetic gold from Sardis,
Get rich otherwise than by my consent to bury him.
Teiresias, it is a sorry thing when a wise man
Sells his wisdom, lets out his words for hire!

TEIRESIAS: Ah Creon! Is there no man left in the world—

CREON: To do what?—Come, let's have the aphorism!

TEIRESIAS: No man who knows that wisdom outweighs any wealth?

CREON: As surely as bribes are baser than any baseness.

TEIRESIAS: You are sick, Creon! You are deathly sick!

CREON: As you say: it is not my place to challenge a prophet.

TEIRESIAS: Yet you have said my prophecy is for sale.

CREON: The generation of prophets has always loved gold.

TEIRESIAS: The generation of kings has always loved brass.

CREON: You forget yourself! You are speaking to your King.

TEIRESIAS: I know it. You are a king because of me.

CREON: You have a certain skill; but you have sold out.

TEIRESIAS: King, you will drive me to words that—

CREON: Say them, say them!
Only remember: I will not pay you for them.

TEIRESIAS: No, you will find them too costly.

CREON: No doubt. Speak:
Whatever you say, you will not change my will.

TEIRESIAS: Then take this, and take it to heart!
The time is not far off when you shall pay back
Corpse for corpse, flesh of your own flesh.
You have thrust the child of this world into living night,
You have kept from the gods below the child that is theirs:
The one in a grave before her death, the other,
Dead, denied the grave. This is your crime:
And the Furies and the dark gods of Hell
Are swift with terrible punishment for you.

Do you want to buy me now, Creon?
Not many days,
And your house will be full of men and women weeping,
And curses will be hurled at you from far
Cities grieving for sons unburied, left to rot
Before the walls of Thebes.

These are my arrows, Creon: they are all for you. (*to* BOY)
But come, child: lead me home.
Let him waste his fine anger upon younger men.
Maybe he will learn at last
To control a wiser tongue in a better head.

(*Exit* TEIRESIAS.)

CHORAGOS: The old man has gone, King, but his words
Remain to plague us. I am old, too,
But I cannot remember that he was ever false.

CREON: That is true. . . . It troubles me.
Oh it is hard to give in! but it is worse
To risk everything for stubborn pride.

CHORAGOS: Creon: take my advice.

CREON: What shall I do?

CHORAGOS: Go quickly: free Antigone from her vault
And build a tomb for the body of Polyneices.

CREON: You would have me do this?

CHORAGOS: Creon, yes!
And it must be done at once: God moves
Swiftly to cancel the folly of stubborn men.

CREON: It is hard to deny the heart! But I
Will do it: I will not fight with destiny.

CHORAGOS: You must go yourself, you cannot leave it to others.

CREON: I will go.
—Bring axes, servants:
Come with me to the tomb. I buried her, I
Will set her free.
Oh quickly!
My mind misgives—
The laws of the gods are mighty, and a man must serve them
To the last day of his life! (*Exit* CREON.)

PAEAN

CHORAGOS: God of many names

CHORUS: O Iacchos
son

of Cadmeian Semele
O born of the Thunder!
Guardian of the West
Regent
of Eleusis' plain
O Prince of maenad Thebes
and the Dragon Field by rippling Ismenos:

CHORAGOS: God of many names

CHORUS: the flame of torches
flares on our hills
the nymphs of Iacchos
dance at the spring of Castalia:

from the vine-close mountain
come ah come in ivy:
Evohe evohe! sings through the streets of Thebes

CHORAGOS: God of many names

CHORUS: Iacchos of Thebes
heavenly Child
of Semele bride of the Thunderer!
The shadow of plague is upon us:
come
with clement feet
oh come from Parnasos
down the long slopes
across the lamenting water

CHORAGOS: Io Fire! Chorister of the throbbing stars!
O purest among the voices of the night!
Thou son of God, blaze for us!

CHORUS: Come with choric rapture of circling Maenads
Who cry to *Io Iacche!*
God of many names!

EXODOS

(*Enter* MESSENGER.)

MESSENGER: Men of the line of Cadmos, you who live
Near Amphion's citadel:
I cannot say
Of any condition of human life "This is fixed,
This is clearly good, or bad." Fate raises up,
And Fate casts down the happy and unhappy alike:
No man can foretell his Fate.
Take the case of Creon:
Creon was happy once, as I count happiness:
Victorious in battle, sole governor of the land,
Fortunate father of children nobly born.
And now it has all gone from him! Who can say
That a man is still alive when his life's joy fails?
He is a walking dead man. Grant him rich,
Let him live like a king in his great house:
If his pleasure is gone, I would not give
So much as the shadow of smoke for all he owns.

CHORAGOS: Your words hint at sorrow: what is your news for us?

MESSENGER: They are dead. The living are guilty of their death.

CHORAGOS: Who is guilty? Who is dead? Speak!

MESSENGER: Haimon.
Haimon is dead; and the hand that killed him
Is his own hand.

CHORAGOS: His father's? or his own?

MESSENGER: His own, driven mad by the murder his father had done.

CHORAGOS: Teiresias, Teiresias, how clearly you saw it all!

MESSENGER: This is my news: you must draw what conclusions you can from it.

CHORAGOS: But look: Eurydice, our Queen:
Has she overheard us?

(*Enter* EURYDICE *from the Palace.*)

EURYDICE: I have heard something, friends:
As I was unlocking the gate of Pallas' shrine,
For I needed her help today, I heard a voice
Telling of some new sorrow. And I fainted
There at the temple with all my maidens about me.
But speak again: whatever it is, I can bear it:
Grief and I are no strangers.

MESSENGER: Dearest Lady,
I will tell you plainly all that I have seen.
I shall not try to comfort you: what is the use,
Since comfort could lie only in what is not true?
The truth is always best.
I went with Creon
To the outer plain where Polyneices was lying,
No friend to pity him, his body shredded by dogs.
We made our prayers in that place to Hecate
And Pluto, that they would be merciful. And we bathed
The corpse with holy water, and we brought
Fresh-broken branches to burn what was left of it,
And upon the urn we heaped up a towering barrow
Of the earth of his own land.
When we were done, we ran
To the vault where Antigone lay on her couch of stone.
One of the servants had gone ahead,
And while he was yet far off he heard a voice
Grieving within the chamber, and he came back
And told Creon. And as the King went closer,

The air was full of wailing, the words lost,
And he begged us to make all haste. "Am I a prophet?"
He said, weeping, "And must I walk this road,
The saddest of all that I have gone before?
My son's voice calls on me. Oh quickly, quickly!
Look through the crevice there, and tell me
If it is Haimon, or some deception of the gods!"

We obeyed; and in the cavern's farthest corner
We saw her lying:
She had made a noose of her fine linen veil
And hanged herself. Haimon lay beside her,
His arms about her waist, lamenting her,
His love lost under ground, crying out
That his father had stolen her away from him.

When Creon saw him the tears rushed to his eyes
And he called to him: "What have you done, child?
Speak to me.
What are you thinking that makes your eyes so strange?
O my son, my son, I come to you on my knees!"
But Haimon spat in his face. He said not a word,
Staring—
And suddenly drew his sword
And lunged. Creon shrank back, the blade missed; and the boy,
Desperate against himself, drove it half its length
Into his own side, and fell. And as he died
He gathered Antigone close in his arms again,
Choking, his blood bright red on her white cheek.
And now he lies dead with the dead, and she is his
At last, his bride in the houses of the dead.

(*Exit* EURYDICE *into the Palace.*)

CHORAGOS: She has left us without a word. What can this mean?

MESSENGER: It troubles me, too; yet she knows what is best,

Her grief is too great for public lamentation,
And doubtless she has gone to her chamber to weep
For her dead son, leading her maidens in his dirge.

CHORAGOS: It may be so: but I fear this deep silence.

MESSENGER: I will see what she is doing. I will go in.

(*Exit* MESSENGER *into the Palace.*)

(*Enter* CREON *with attendants, bearing* HAIMON*'s body.*)

CHORAGOS: But here is the King himself: oh look at him,
Bearing his own damnation in his arms.

CREON: Nothing you say can touch me any more.
My own blind heart has brought me
From darkness to final darkness. Here you see
The father murdering, the murdered son—
And all my civic wisdom!
Haimon my son, so young, so young to die,
I was the fool, not you; and you died for me.

CHORAGOS: That is the truth; but you were late in learning it.

CREON: This truth is hard to bear. Surely a god
Has crushed me beneath the hugest weight of heaven,
And driven me headlong a barbaric way
To trample out the thing I held most dear.
The pains that men will take to come to pain!

(*Enter* MESSENGER *from the Palace.*)

MESSENGER: The burden you carry in your hands is heavy,
But it is not all: you will find more in your house.

CREON: What burden worse than this shall I find there?

MESSENGER: The Queen is dead.

CREON: O port of death, deaf world,
Is there no pity for me? And you, Angel of evil,

I was dead, and your words are death again.
Is it true, boy? Can it be true?
Is my wife dead? Has death bred death?

MESSENGER: You can see for yourself.

(*The doors are opened, and the body of* EURYDICE *is disclosed within.*)

CREON: Oh pity!
All true, all true, and more than I can bear!
O my wife, my son!

MESSENGER: She stood before the altar, and her heart
Welcomed the knife her own hand guided,
And a great cry burst from her lips for Megareus dead,
And for Haimon dead, her sons; and her last breath
Was a curse for their father, the murderer of her sons.
And she fell, and the dark flowed in through her closing eyes.

CREON: O God, I am sick with fear.
Are there no swords here? Has no one a blow for me?

MESSENGER: Her curse is upon you for the deaths of both.

CREON: It is right that it should be. I alone am guilty.
I know it, and I say it. Lead me in,
Quickly, friends.
I have neither life nor substance. Lead me in.

CHORAGOS: You are right, if there can be right in so much wrong.
The briefest way is best in a world of sorrow.

CREON: Let it come,
Let death come quickly, and be kind to me.
I would not ever see the sun again.

CHORAGOS: All that will come when it will; but we, meanwhile,
Have much to do. Leave the future to itself.

CREON: All my heart was in that prayer!

CHORAGOS: Then do not pray any more: the sky is deaf.

CREON: Lead me away. I have been rash and foolish.
I have killed my son and my wife.
I look for comfort; my comfort lies here dead.
Whatever my hands have touched has come to nothing.
Fate has brought all my pride to a thought of dust.

(*As* CREON *is being led into the house, the* CHORAGOS *advances and speaks directly to the audience.*)

CHORAGOS: There is no happiness where there is no wisdom;
No wisdom but in submission to the gods.
Big words are always punished,
And proud men in old age learn to be wise.